NEWS

NASA

NATIONAL AERONAUTICS AND SPACE ADMINISTRATION
WASHINGTON, D.C. 20546

TELS. WO 2-4155
WO 3-6925

FOR RELEASE: IMMEDIATE
May 7, 1969

RELEASE NO: 69-68

PROJECT: APOLLO 10

I0040886

P R E S S   K I T

contents

5/6/69

Published by Books Express Publishing
Copyright © Books Express, 2012
ISBN 978-1-78039-859-4

Books Express publications are available from all good retail and online booksellers. For
publishing proposals and direct ordering please contact us at: info@books-express.com

Contents Continued

NEWS **NASA**

NATIONAL AERONAUTICS AND SPACE ADMINISTRATION
WASHINGTON, D.C. 20546

TELS. WO 2-4155
WO 3-6925

FOR RELEASE: IMMEDIATE
May 7, 1969

RELEASE NO: 69-68

APOLLO 10:   MAN'S NEAREST LUNAR APPROACH

Two Apollo 10 astronauts will descend to within eight nautical miles of the Moon's surface, the closest man has ever been to another celestial body.

A dress rehearsal for the first manned lunar landing, Apollo 10 is scheduled for launch May 18 at 12:49 p.m. EDT from the National Aeronautics and Space Administration's Kennedy Space Center, Fla.

The eight-day, lunar orbit mission will mark the first time the complete Apollo spacecraft has operated around the Moon and the second manned flight for the lunar module.

Following closely the time line and trajectory to be flown on Apollo 11, Apollo 10 will include an eight-hour sequence of lunar module (LM) undocked activities during which the commander and LM pilot will descend to within eight nautical miles of the lunar surface and later rejoin the command/service module (CSM) in a 60-nautical-mile circular orbit.

-more-

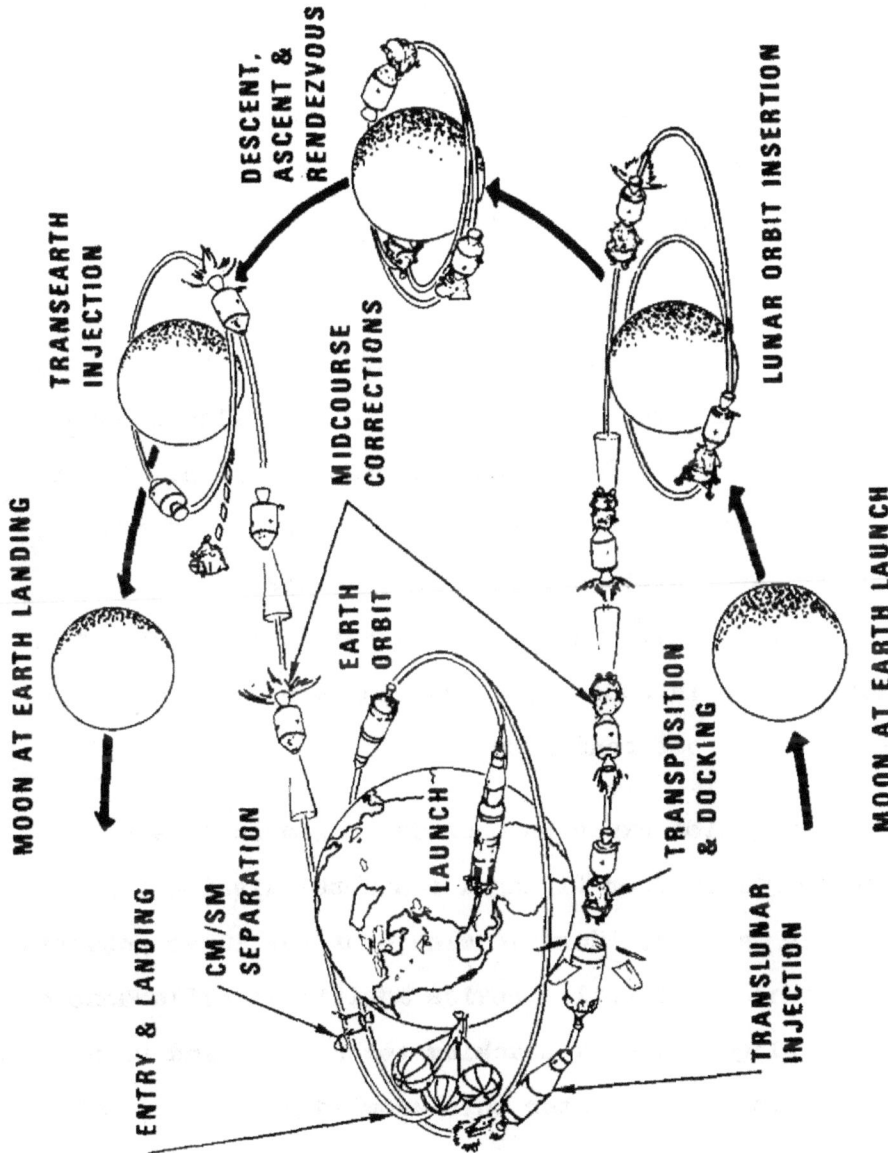

APOLLO LUNAR MISSION

All aspects of Apollo 10 will duplicate conditions of the lunar landing mission as closely as possible--Sun angles at Apollo Site 2, the out-and-back flight path to the Moon, and the time line of mission events. Apollo 10 differs from Apollo 11 in that no landing will be made on the Moon's surface.

Apollo 10 is designed to provide additional operational experience for the crew; space vehicle; and mission-support facilities during a simulated lunar landing mission. Among desired data points to be gained by Apollo 10 are LM systems operations at lunar distances as well as overall mission operational experience. The LM was successfully checked-out in Earth orbit in Apollo 9, including a rendezvous sequence simulating lunar orbit rendezvous.

Space navigation experience around the Moon is another benefit to be gained from flying a rehearsal mission before making a lunar landing. More knowledge of the lunar potential, or gravitational effect will provide additional refinement of Manned Space Flight Network tracking techniques, and broad landmark tracking will bolster this knowledge.

Analysis of last December's Apollo 8 lunar orbit mission tracking has aided refinement of tracking and navigation techniques and Apollo 10 should reduce error margins still further.

Apollo 10 crewmen are Commander Thomas P. Stafford, Command Moudle Pilot John W. Young and Lunar Module Pilot Eugene A. Cernan. The mission will be the third space flight for Stafford (Gemini 6 and 9) and Young (Gemini 3 and 10), and the second for Cernan (Gemini 9). The three were recycled from the Apollo 7 backup crew. The Apollo 10 backup crew is Commander L. Gordon Cooper, Command Moudle Pilot Donn F. Eisele and Lunar Module Pilot Edgar D. Mitchell.

Stafford is an Air Force Colonel; Young and Cernan are Navy Commanders.

If necessary, the backup crew can be substituted for the prime crew up to about two weeks prior to an Apollo launch. During this period, the flight hardware and software, ground hardware and software, flight crew and ground crews work as an integrated team to perform ground simulations and other tests of the upcoming mission. It is necessary that the flight crew that will conduct the mission take part in these activities, which are not repeated for the benefit of the backup crew. To do so would add an additional costly two-week period to the pre-launch schedule, which, for a lunar mission, would require rescheduling for the next lunar window.

The Apollo 10 rendezvous will be the fifth space rendezvous in which Stafford has taken part--Gemini 7/6 and the world's first rendezvous, and three types of rendezvous with the augmented target docking adapter in Gemini 9.

The Apollo 10 mission time line can be described as a combination of Apollo 8 and Apollo 9 in that it will be a lunar orbit mission with a CSM-LM rendezvous. Apollo 8 was a lunar orbit mission with the command/service module only, while Apollo 9 was an Earth orbital mission with the complete Apollo spacecraft and included a LM-active rendezvous with the CSM.

Apollo 10, after liftoff from Launch Complex 39B, will begin the three-day voyage to the Moon about two and a half hours after the spacecraft is inserted into a 100-nautical mile circular Earth parking orbit. The Saturn V launch vehicle third stage will restart to inject Apollo 10 into a translunar trajectory as the vehicle passes over Australia mid-way through the second revolution of the Earth.

The "go" for translunar injection will follow a complete checkout of the spacecraft's readiness to be committed for injection. About an hour after translunar injection (TLI), the command/service module will separate from the Saturn third stage, turn around and dock with the lunar module nested in the spacecraft LM adapter. Spring-loaded lunar module holddowns will be released to eject the docked spacecraft from the adapter.

Later, leftover liquid propellant in the Saturn third stage will be vented through the engine bell to place the stage into a "slingshot" trajectory to miss the Moon and go into solar orbit.

During the translunar coast, Apollo 10 will be in the so-called passive thermal control mode in which the space-craft rotates slowly about one of its axes to stabilize thermal response to solar heating. Four midcourse correction maneuvers are possible during translunar coast and will be planned in real time to adjust the trajectory.

Apollo 10 will first be inserted into a 60-by-170-nautical mile elliptical lunar orbit, which two revolutions later will be circularized to 60 nautical miles. Both lunar orbit inser-tion burns (LOI) will be made when Apollo 10 is behind the Moon out of "sight" of Manned Space Flight Network stations.

Stafford and Cernan will man the LM for systems checkout and preparations for an eight-and-a-half hour sequence that duplicates--except for an actual landing--the maneuvers planned for Apollo 11. The LM twice will sweep within 50,000 feet of Apollo Landing Site 2, one of the prime targets for the Apollo 11 landing.

Maximum separation between the LM and the CSM during the rendezvous sequence will be about 350 miles and will provide an extensive checkout of the LM rendezvous radar as well as of the backup VHF ranging device aboard the CSM, flown for the first time on Apollo 10.

When the LM ascent stage has docked with the CSM and the two crewmen have transferred back to the CSM, the LM will be jettisoned for a ground command ascent engine burn to pro-pellant depletion which will place the LM ascent stage into solar oribt.

The crew of Apollo 10 will spend the remainder of the time in lunar oribt conducting lunar navigational tasks and photographing Apollo landing sites that are within camera range of Apollo 10's ground track.

The transearth injection burn will be made behind the Moon after 61.5 hours in lunar orbit. During the 54-hour transearth coast, Apollo 10 again will control solar heat loads by using the passive thermal control "barbecue" tech-nique. Three transearth midcourse corrections are possible and will be planned in real time to adjust the Earth entry corridor.

Apollo 10 will enter the Earth's atmosphere (400,000 feet) at 191 hours 51 minutes after launch at 36,310 feet-per-second. Command module touchdown will be 1,285 nautical miles down-range from entry at 15 degrees 7 minutes South latitude by 165 degrees West longitude at an elapsed time of 192 hours 5 minutes. The touchdown point is about 345 nautical miles east of Pago Pago, Tutuila, in American Samoa.

(END OF GENERAL RELEASE; BACKGROUND INFORMATION FOLLOWS)

APOLLO 10 — Launch and Trans Lunar Injection

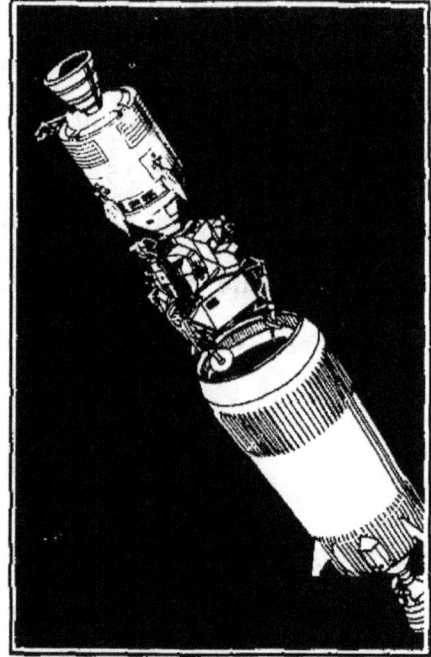

Saturn Staging

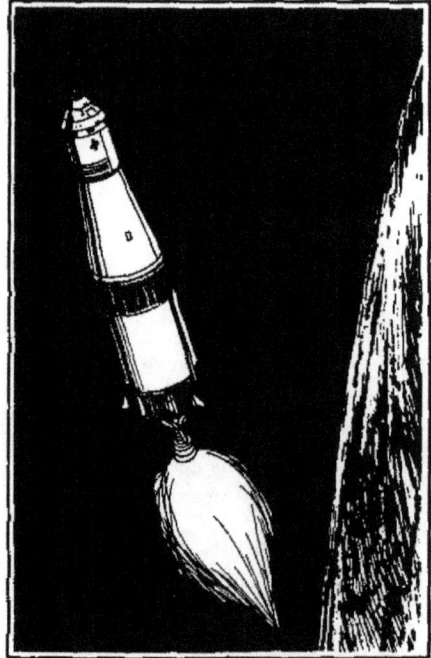

Apollo Saturn Separation

Astronauts Board Apollo

Trans Lunar Injection

APOLLO 10—Trans Lunar Flight

Apollo Midcourse Maneuver

Astronauts at Command Module Stations

APOLLO 10 — Trans Lunar Flight

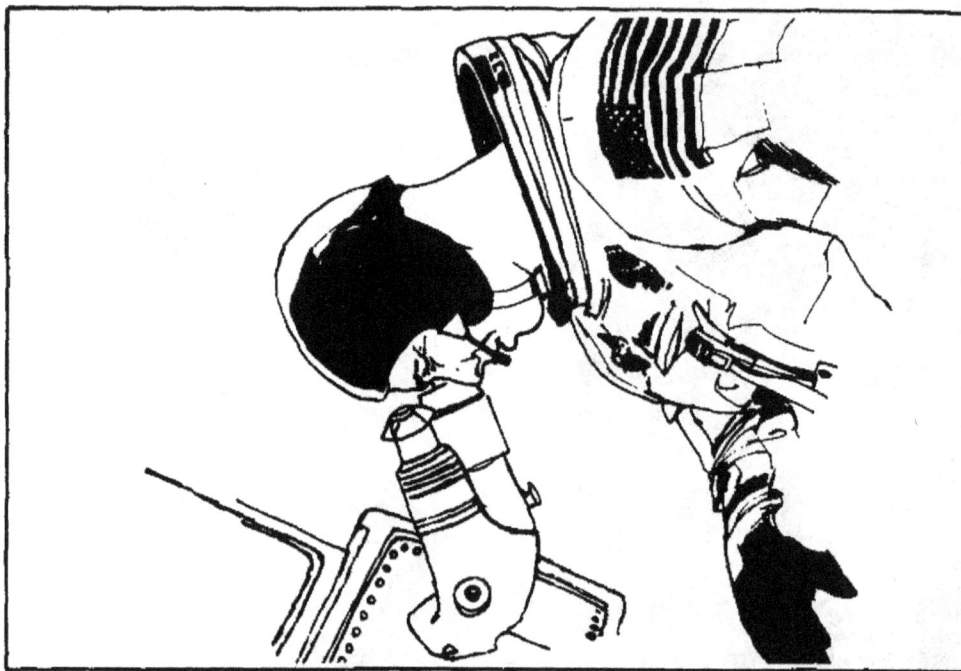

Navigational Check

Final Course Adjustment

APOLLO 10 — Lunar Orbital Flight

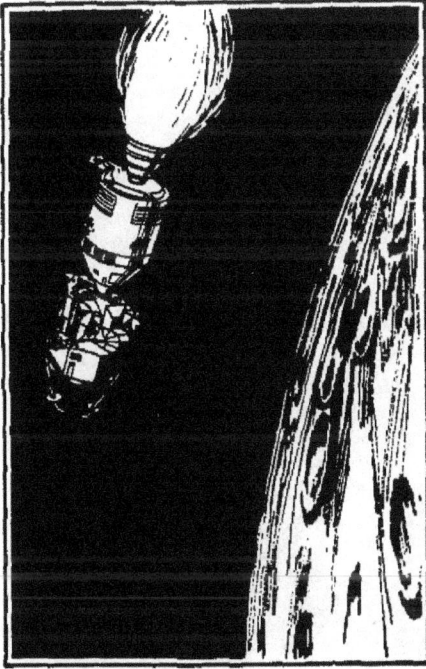

Television Broadcast

Transfer to Lunar Module

Lunar Orbit Insertion

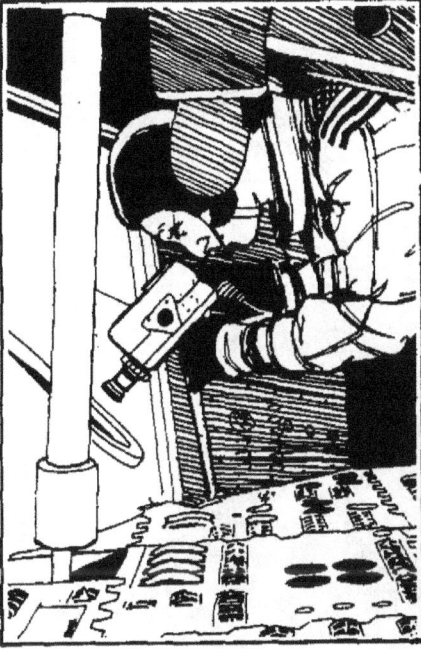

Lunar Landmark Tracking

APOLLO 10 — Lunar Descent and Rendezvous

Descent Orbit Insertion   Lunar Module Staging   Apollo Docking

LM
Ascent Engine Firing to Depletion

Lunar Landmark Tracking

APOLLO 10 — Trans Earth Injection and Flight

Apollo Midcourse Maneuver

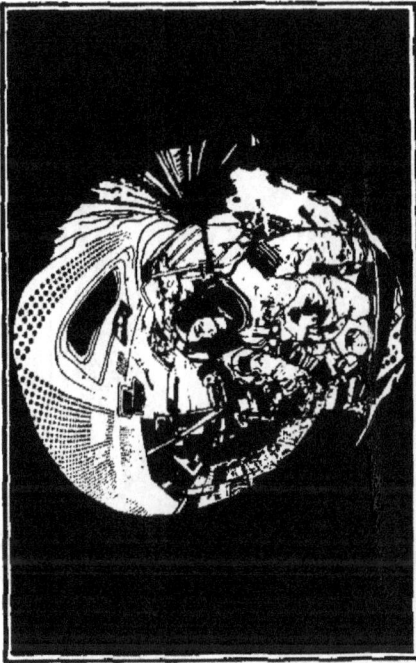

Final Reentry Preparations

Trans Earth Injection

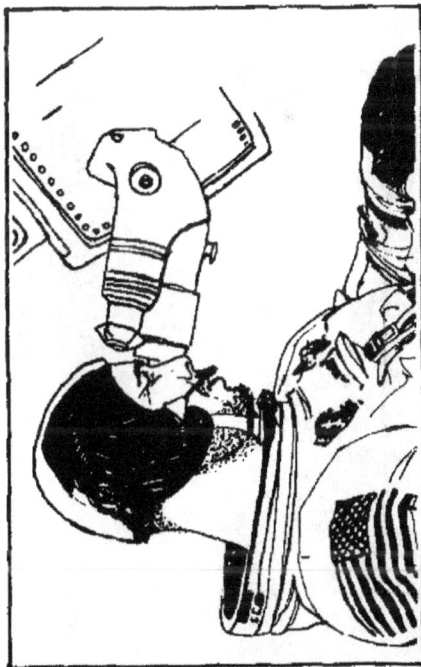

Navigational Check

APOLLO 10 — Earth Reentry and Recovery

Command Module Reentry

Recovery

Command-Service Module Separation

Splashdown

MISSION OBJECTIVES

Although Apollo 10 will pass no closer than eight
nautical miles from the lunar surface, all other aspects
of the mission will be similar to the first lunar landing
mission, Apollo 11, now scheduled for July.

The trajectory, time line and maneuvers follow the
lunar landing profile. After rendezvous is completed, the
Apollo 10 time line will deviate from Apollo 11 in that
Apollo 10 will spend an extra day in lunar orbit.

Additional LM operation in either Earth orbit or lunar
orbit will provide additional experience and confidence with
the LM systems, including various control modes of the LM
primary/abort guidance systems, as well as further assessment
of crew time lines.

The mission will also test the Apollo rendezvous radar
at maximum range (approximately 350 miles vs. 100 miles during
Apollo 9). Apollo 10 will mark the first space flight test
of the LM steerable S-band antenna and of the LM landing radar.
The LM landing radar has undergone numerous tests in Earth environment, but this mission will provide an opportunity to
check the lunar surface reflectivity characteristics with the
landing radar.

Some 800 seconds of landing radar altitude-measuring
data will be gathered as the LM makes two sweeps eight nautical
miles above Apollo Landing Site 2.

This mission will also provide the first opportunity to
check the very high frequency (VHF) ranging device aboard the
CSM which serves as a backup to the LM rendezvous radar.

The Apollo 10 mission profile provides fuel and other con-
sumable reserves in the LM that are greater than those planned
for the first LM to land on the Moon. The lunar landing mission
is the "design mission" for the Apollo spacecraft, and such a
mission has smaller although adequate margins of reserve consumables.

From liftoff through descent orbit insertion, Apollo 10
follows closely the trajectory and time line that will be flown
in the landing mission. Following the eight-mile pericynthion,
the profile closely simulates the conditions of lunar orbit
rendezvous after a landing.

The May 18 launch date will produce lighting conditions on Apollo Site 2 similar to those that will be present for the landing mission. At the low inclination to be flown on Apollos 10 and 11 -- about 1.2 degrees relative to the lunar equator-- Apollo landing Site 3 can be photographed and optically tracked by the crew of Apollo 10 in addition to the prime Site 2.

Site 1 was photographed by Apollo 8 in last December's lunar orbit mission and, together with the two sites to be covered in Apollo 10, photographic, tracking and site altitude data on three sites will be in hand.

Among the Apollo 10 objectives is the gathering of additional Manned Space Flight Network (MSFN) tracking data on vehicles in lunar orbit. While MSFN experience in tracking Apollo 8 will benefit Apollo 10, there are still some uncertainties. For example, there is still some lack of knowledge as to what the exact lunar potential or gravity field is and how it affects an orbiting spacecraft.

In tracking Apollo 8, downtrack, or orbital timing errors projected ahead two revolutions were 30,000 feet, and orbital radius measurements relative to the center of the Moon were off 5,500 feet. MSFN tracking can produce accurate position and velocity information in real time while a spacecraft is "in view" from the Earth and not occulted by the Moon, but landing and rendezvous operations will require accurate predictions of position and velocity several revolutions in advance of the event.

The lunar potential apparently affects an orbiting spacecraft differently depending upon orbital inclination and altitude. Apollo 10 will be flown on the same inclination to the lunar equator as the landing mission and will provide information for refining prediction techniques.

Apollo 8 postflight analysis has produced modifications to tracking and position prediction techniques which should reduce downtrack errors to 3,000 feet and altitude errors to 1,400 feet. Apollo 10 will allow mission planners to perfect techniques developed as a result of Apollo 8 tracking analysis.

Other space navigation benefits from Apollo 10 will be gained from combining onboard spacecraft lunar landmark tracking data with MSFN tracking and from evaluating present lunar landing site maps at close visual and camera ranges.

Additionally, LM descent and ascent engine burns will be monitored by MSFN stations for developing useful techniques for tracking powered flight in future missions.

APOLLO 10 COUNTDOWN

The clock for the Apollo 10 countdown will start at T-28 hours, with a six hour built-in-hold planned at T-9 hours, prior to launch vehicle propellant loading.

The countdown is preceded by a pre-count operation that begins some 4 days before launch. During this period the tasks include mechanical buildup of both the command/service module and LM, fuel cell activation and servicing and loading of the super critical helium aboard the LM descent stage. A 5½ hour built-in-hold is scheduled between the end of the pre-count and start of the final countdown.

Following are some of the highlights of the final count:

T-28 hrs.	Official countdown starts
T-27 hrs. 30 mins.	Install launch vehicle flight batteries (to 23 hrs. 30 mins.) LM stowage and cabin closeout (to 15 hrs.)
T-21 hrs.	Top off LM super critical helium (to 19 hrs.)
T-16 hrs.	Launch vehicle range safety checks (to 15 hrs.)
T-11 hrs. 30 mins.	Install launch vehicle destruct devices (to 10 hrs. 45 mins.) Command/service module pre-ingress operations
T-10 hrs.	Start mobile service structure move to park site
T-9 hrs.	Start six hour built-in-hold
T-9 hrs. counting	Clear blast area for propellant loading
T-8 hrs. 30 mins.	Astronaut backup crew to spacecraft for prelaunch checks
T-8 hrs. 15 mins.	Launch Vehicle propellant loading, three stages (liquid oxygen in first stage) liquid oxygen and liquid hydrogen in second, third stages. Continues thru T-3 hrs. 38 mins.

-more-

T-5 hrs.	Flight crew alerted
T-4 hrs. 45 mins.	Medical examination
T-4 hrs. 15 mins.	Breakfast
T-3 hrs. 45 mins.	Don space suits
T-3 hrs. 30 mins.	Depart Manned Spacecraft Operations Building for LC-39 via crew transfer van
T-3 hrs. 14 mins.	Arrive at LC-39
T-3 hrs. 10 mins.	Enter Elevator to spacecraft level
T-2 hrs. 40 mins.	Start flight crew ingress
T-1 hr. 55 mins.	Mission Control Center-Houston/spacecraft command checks
T-1 hr. 50 mins.	Abort advisory system checks
T-1 hr. 46 mins.	Space vehicle Emergency Detection System (EDS) test
T-43 mins.	Retract Apollo access arm to standby position (12 degrees)
T-42 mins.	Arm launch escape system
T-40 mins. '	Final launch vehicle range safety checks (to 35 mins.)
T-30 mins.	Launch vehicle power transfer test
	LM switch over to internal power
T-20 mins. to T-10 mins.	Shutdown LM operational instrumentation
T-15 mins.	Spacecraft to internal power
T-6 mins.	Space vehicle final status checks
T-5 mins. 30 sec.	Arm destruct system
T-5 mins.	Apollo access arm fully retracted
T-3 mins. 10 sec.	Initiate firing command (automatic sequencer)
T-50 sec.	Launch vehicle transfer to internal power

T-8.9 sec.	Ignition sequence start
T-2 sec.	All engines running
T-0	Liftoff

*Note: Some changes in the above countdown are possible as a result of experience gained in the Countdown Demonstration Test (CDDT) which occurs about 10 days before launch.

MISSION TRAJECTORY AND MANEUVER DESCRIPTION

(Note: Information presented herein is based upon a
May 18 launch and is subject to change prior to the mission
or in real time during the mission to meet changing conditions.)

Launch

Apollo 10 will be launched from Kennedy Space Center Launch
Complex 39B on a launch azimuth that can vary from 72 degrees to
108 degrees, depending upon the time of day of launch. The
azimuth changes with time of day to permit a fuel-optimum injection
from Earth parking orbit into a free-return circumlunar trajectory.
Other factors influencing the launch windows are a daylight launch
and proper Sun angles on lunar landing sites.

The planned Apollo 10 launch date of May 18 will call for
liftoff at 12:49 p.m. EDT on a launch azimuth of 72 degrees.
Insertion into a 100-nautical-mile circular Earth parking orbit
will occur at 11 minutes 53 seconds ground elapsed from launch
(GET), and the resultant orbit will be inclined 32.5 degrees to
the Earth's equator.

FLIGHT PROFILE

TRANSEARTH INJECTION BURN

CSI 45 N.MI.

TPF 60 N.MI.

LM PHASING ORBIT 9x194 N.MI.

CSM 60 N.MI.

LM DESCENT

LUNAR ORBIT INSERTION

CIRCULARIZATION

TPI

DOCKING

CDH

CSM/LM SEPARATION

LM INSERTION BURN 9x45 N.MI

LM PHASING BURN

CSM TRANSEARTH TRAJECTORY

CSM 60 N.MI.

50,000 FT.

60x170 N.MI. LUNAR ORBIT

60 N.M. LUNAR ORBIT

CM/SM SEPARATION

100 N.MI. EARTH PARKING ORBIT

INSERTION

LAUNCH

S-IVB RESIDUAL PROPELLANT DUMP (SLINGSHOT)

EARTH

CM

S-IVB RESTART DURING 2ND OR 3RD ORBIT

CM SPLASHDOWN & RECOVERY

S-IVB 2ND BURN CUTOFF TRANSLUNAR INJECTION

S/C SEPARATION, TRANSPOSITION, DOCKING & EJECTION

SPACE VEHICLE LAUNCH EVENTS/WEIGHTS

Time Hrs.	Min.	Sec.	Event	Altitude Naut. Mi.	Velocity Knots	Weight Pounds
00	00	(-)08.9	Ignition	0.00	0	6,499,016
00	00	00	First Motion	0.033	*0	6,412,918
00	00	12	Tilt Initiation	0.12	*3	--
00	01	21	Maximum Dynamic Pressure	7	1554	--
00	02	15	Center Engine Cutoff	24	3888	2,434,985
00	02	40	Outboard Engines Cutoff	35	5324	1,842,997
00	02	41	S-IC/S-II Separation	36	5343	1,465,702
00	02	42	S-II Ignition	37	5335	1,465,123
00	03	11	S-II Aft Interstage Jettison	49	5581	--
00	03	16	LES Jettison	51	5642	--
00	03	21	Initiate IGM	53	5701	
00	07	39	S-II Center Engine Cutoff	97	10977	644,128
00	09	14	S-II Outboard Engines Cutoff	102	13427	471,494
00	09	15	S-II/S-IVB Separation	102	13434	364,429
00	09	18	S-IVB Ignition	102	13434	364,343
00	11	43	S-IVB First Cutoff	103	15135	295,153
00	11	53	Parking Orbit Insertion	103	15139	295,008

*First two velocities are space fixed. Others are inertial velocities. Vehicle on launch pad has inertial velocity of 408.5 meters per second (793.7 knots).

The above figures are based on a launch azimuth of 72 degrees. Figures will vary slightly for other azimuths.

Apollo 10 Mission Events

Event	Ground Elapsed Time hrs:min:sec	Date & Time (EDT)	Velocity Change feet/sec	Purpose and (Resultant Orbit)
Insertion	00:11:53	5/18 1:01 pm	25,593	Insertion into 100 nm circular EPO.
Translunar injection	02:33:26	5/18 3:23 pm	10,058	Injection into free-return translunar trajectory with 60 nm pericynthion.
CSM separation, docking	03:10:00	5/18 3:59 pm	--	Hard-mating of CSM and LM.
Ejection from SLA	04:09:00	5/18 4:58 pm	1	Separates CSM-LM from S-IVB/SLA.
SPS evasive maneuver	04:29:00	5/18 5:18 pm	19.7	Provides separation prior to S-IVB propellant dump and "slingshot" maneuver.
Midcourse correction No. 1	TLI +9 hrs.	5/19 12:22 am	55	* These midcourse corrections have a nominal velocity change of 0 fps, but will be calculated in real time to correct TLI dispersions. MCC-3 will have a plane change component to achieve desired lunar orbit inclination.
Midcourse correction No. 2	TLI +24 hrs.	5/19 3:23 pm	0	
Midcourse correction No. 3	LOI -22 hrs.	5/20 6:35 pm	0	
Midcourse correction No. 4	LOI -5 hrs.	5/21 11:35 am	0	
Lunar Orbit Insertion No. 1	75:45:43	5/21 4:35 pm	-2,974	Inserts Apollo 10 into 60x170 nm elliptical lunar orbit.
Lunar Orbit Insertion No. 2	80:10:45	5/21 9:00 pm	-138.5	Circularizes lunar parking orbit to 60 nm.
CSM-LM undocking; separation (SM RCS)	98:10:00 98:35:16	5/22 2:59 pm 5/22 3:24 pm	2.5	Establishes equiperiod orbit for 2 nm separation (minifootball)
Descent orbit insertion (DPS)	99:33:59	5/22 4:23 pm	-71	Lower LM pericynthion to eight nm (8x60).

~more~

Event	Ground Elapsed Time hrs:min:sec	Date & Time (EDT)	Velocity Change feet/sec	Purpose and (Resultant Orbit)
DPS phasing burn	100:46:21	5/22 5:35 pm	195	Raises LM apocynthion to 194 nm, allows CSM to pass and overtake LM (8x194).
APS insertion burn	102:43:18	5/22 7:32 pm	-207	Simulates LM ascent into lunar orbit after landing (8x43.6).
LM RCS concentric sequence initiate (CSI) burn	103:33:46	5/22 8:22 pm	50.5	Raises LM pericynthion to 46.2 nm, adjusts orbital shape for rendezvous sequence (42.9x46.2).
LM RCS constant delta height (CDH) burn	104:31:42	5/22 9:20 pm	3.4	Radially downward burn adjusts LM to constant 15 nm below CSM.
LM RCS terminal phase initiate (TPI) burn	105:09:00	5/22 9:58 pm	24.6	LM thrusts along line-of-sight toward CSM, midcourse and braking maneuvers as necessary.
Rendezvous (TPF)	105:54:00	5/22 10:43 pm	--	Completes rendezvous sequence. Fly formation at 100 ft.
Docking	106:20:00	5/22 11:09 pm	--	Transfer back to CSM (about 107 GET).
APS burn to depletion	108:38:57	5/23 1:28 am	3,837	Posigrade APS depletion burn near LM pericynthion injects LM ascent stage into heliocentric orbit.
Transearth injection (TEI) SPS burn	137:20:22	5/24 2:09 am	3,622.5	Injects CSM into 54½-hour transearth trajectory.

-more-

Event	Ground Elapsed Time hrs:min:sec	Date & Time (EDT)	Velocity Change feet/sec	Purpose and (Resultant Orbit)
Midcourse correction No. 5	TEI +15 hrs.	5/24 5:09 pm	--	• Transearth midcourse corrections will be computed in real time for entry corridor control and for adjusting landing point to avoid recovery area foul weather.
Midcourse correction No. 6	Entry - 15 hrs.	5/25 5:39 pm	--	
Midcourse correction No. 7	Entry - 3 hrs.	5/26 5:39 am	--	
CM/SM separation	191:35	5/26 8:24 am	--	Reentry condition.
Entry interface (400,000 feet)	191:50:32	5/26 8:39 am	--	Command module enters Earth's sensible atmosphere at 36,310 fps.
Touchdown	192:04:47	5/26 8:54 am	--	Landing 1,285 nm downrange from entry 15 degrees seven minutes South latitude x 165 degrees West longitude.

-more-

The crew for the first time will have a backup to launch vehicle guidance during powered flight. If the Saturn instrument unit inertial platform fails, the crew can switch guidance to the command module computer for first-stage powered flight automatic control. Second and third stage backup guidance is through manual takeover in which command module hand controller inputs are fed through the command module computer to the Saturn instrument unit.

Earth Parking Orbit (EPO)

Apollo 10 will remain in Earth parking orbit for one-and-one-half revolutions after insertion and will hold a local horizontal attitude during the entire period. The crew will perform spacecraft systems checks in preparation for the translunar injection (TLI) burn. The final "go" for the TLI burn will be given to the crew through the Carnarvon, Australia, Manned Space Flight Network station.

Translunar Injection (TLI)

Midway through the second revolution in Earth parking orbit, the S-IVB third-stage engine will reignite at two hours 33 minutes 26 seconds Ground Elapsed Time (GET) over Australia to inject Apollo 10 toward the Moon. The velocity will increase from 25,593 feet-per-second (fps) to 35,651 fps at TLI cutoff -- a velocity increase of 10,058 fps. The TLI burn will place the spacecraft on a free-return circumlunar trajectory from which midcourse corrections could be made with the SM reaction control system thruster. Splashdown for a free-return trajectory would be at 6:37 p.m. EDT May 24 at 24.9 degrees South latitude by 84.3 degrees East longitude after a flight time of 149 hours and 49 minutes.

Transposition, Docking and Ejection (TD&E)

At about three hours after liftoff and 25 minutes after the TLI burn, the Apollo 10 crew will separate the command/service module from the spacecraft lunar module adapter (SLA), thrust out away from the S-IVB, turn around and move back in for docking with the lunar module. Docking should take place at about three hours and ten minutes GET, and after the crew confirms all docking latches solidly engaged, they will connect the CSM-to-LM umbilicals and pressurize the LM with the command module surge tank. At about 4:09 GET, docked spacecraft will be ejected from the spacecraft LM adapter by spring devices at the four LM landing gear "knee" attach points. The ejection springs will impart about one fps velocity to the spacecraft. A 19.7 fps service propulsion system (SPS) evasive maneuver in plane at 4:29 GET will separate the spacecraft to a safe distance for the S-IVB "slingshot" maneuver in which residual liquid propellants will be dumped through the J-2 engine bell to propel the stage into a trajectory passing behind the Moon's trailing edge and on into solar orbit.

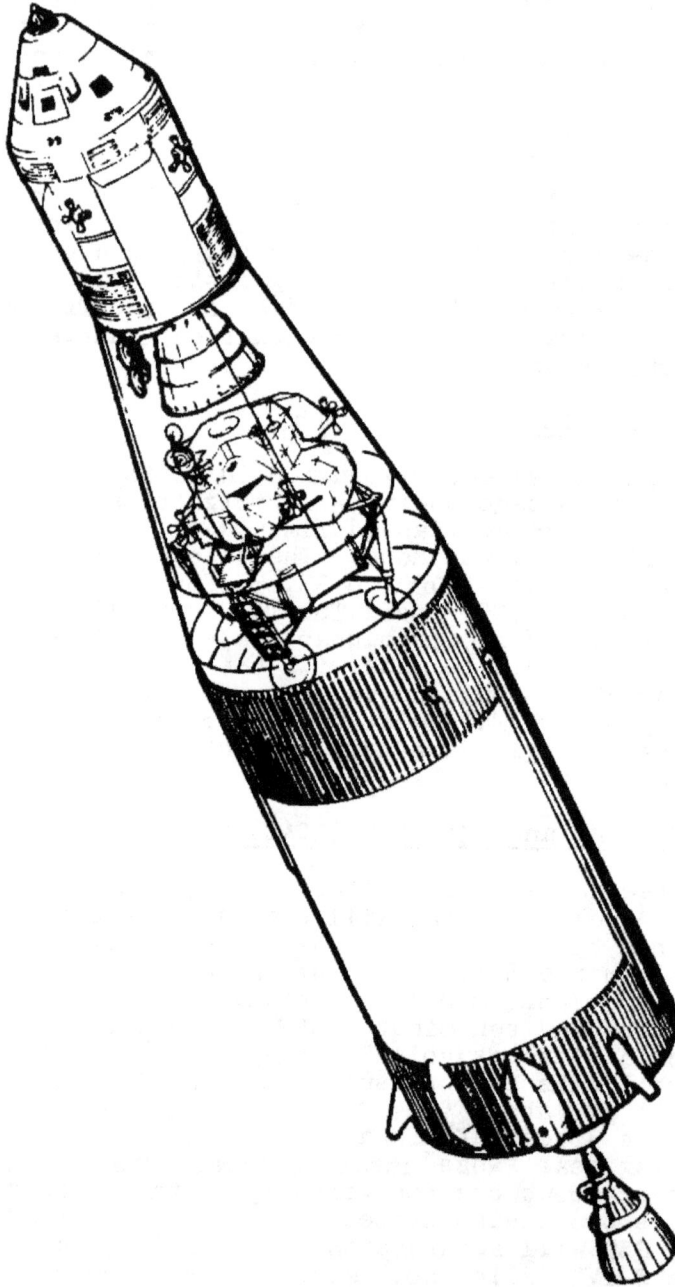

VEHICLE EARTH PARKING ORBIT CONFIGURATION

(SATURN V THIRD STAGE AND INSTRUMENT UNIT, APOLLO SPACECRAFT)

SPACE

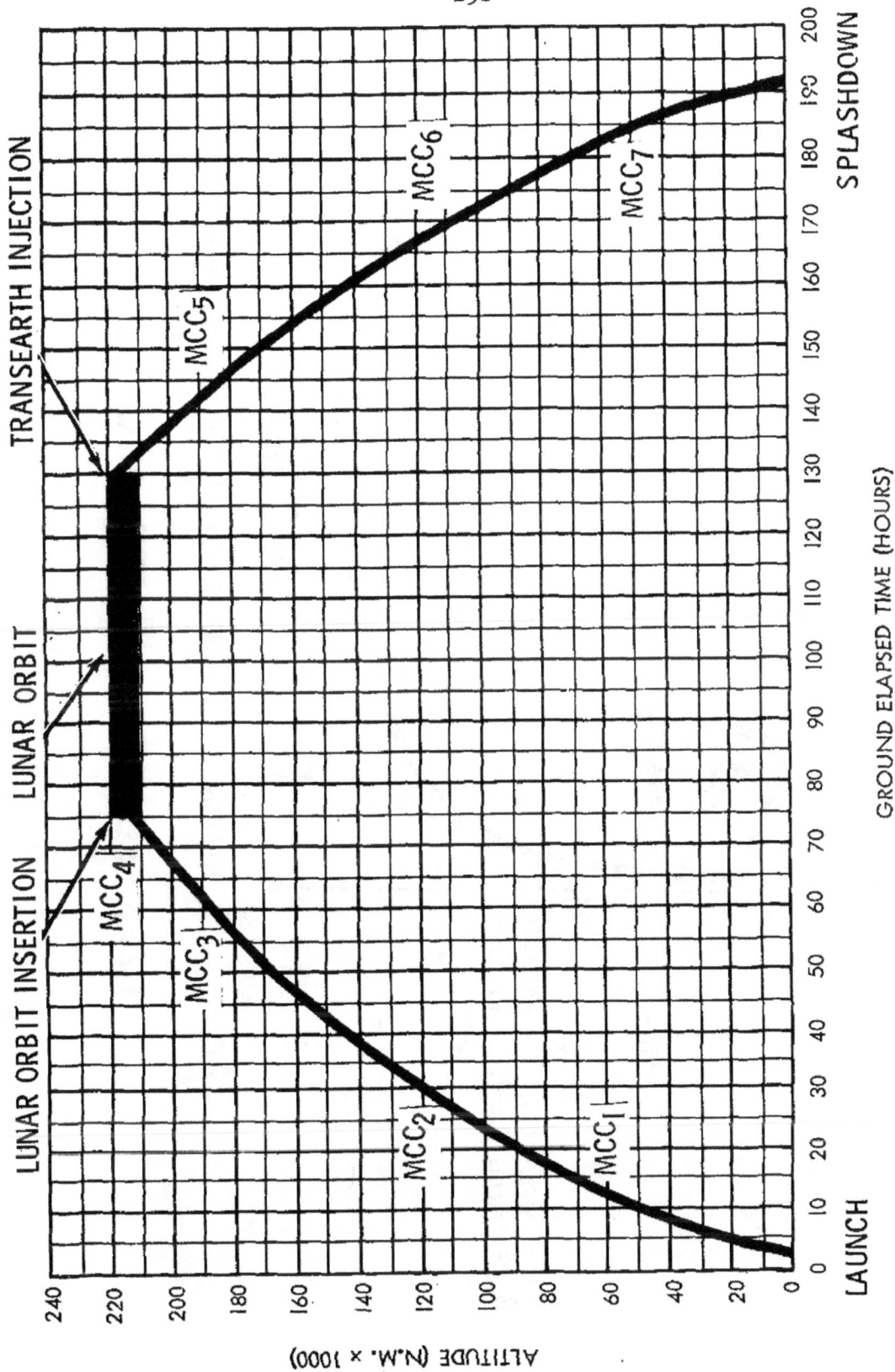

SPACECRAFT ALTITUDE VS. TIME

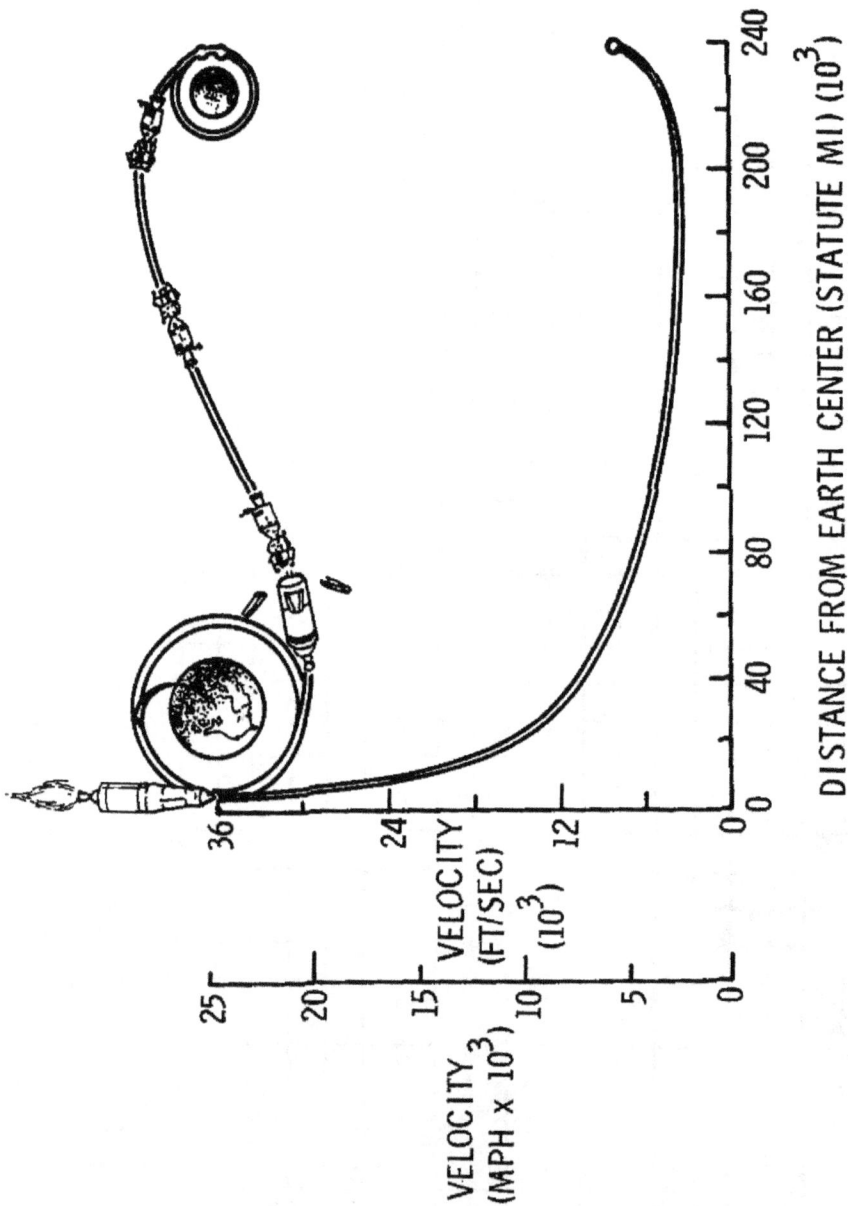

TRANSLUNAR VELOCITY PROFILE

DISTANCE FROM EARTH CENTER (STATUTE MI) (10^3)

VELOCITY (FT/SEC) (10^3)

VELOCITY (MPH x 10^3)

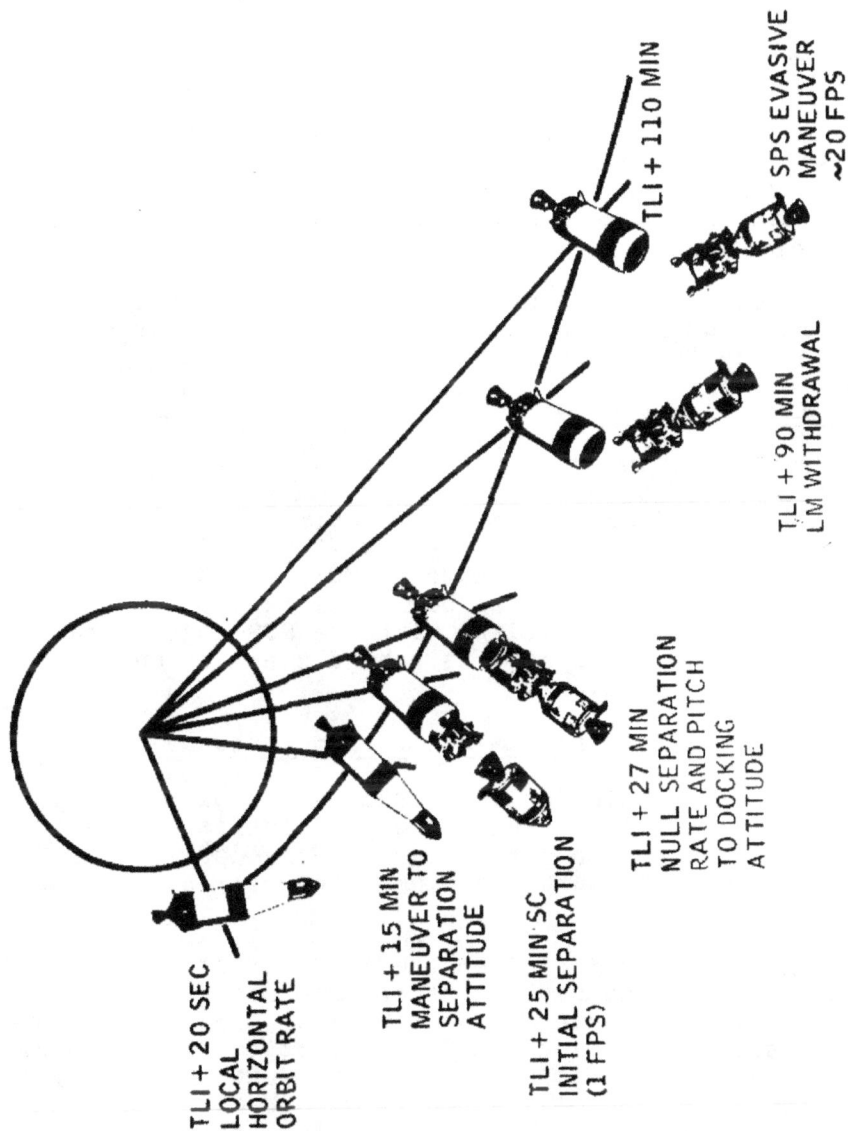

POST TLI TIMELINE

TLI + 110 MIN

SPS EVASIVE MANEUVER ~20 FPS

TLI + 90 MIN LM WITHDRAWAL

TLI + 27 MIN NULL SEPARATION RATE AND PITCH TO DOCKING ATTITUDE

TLI + 25 MIN SC INITIAL SEPARATION (1 FPS)

TLI + 15 MIN MANEUVER TO SEPARATION ATTITUDE

TLI + 20 SEC LOCAL HORIZONTAL ORBIT RATE

Translunar Coast

Up to four midcourse correction burns are planned during the translunar coast phase, depending upon the accuracy of the trajectory resulting from the TLI maneuver. If required, the midcourse correction burns are planned at TLI +9 hours, TLI +24 hours, lunar orbit insertion (LOI) -22 hours and LOI -5 hours.

During coast periods between midcourse corrections, the spacecraft will be in the passive thermal control (PTC) or "barbecue" mode in which the spacecraft will rotateslowly about one axis to stabilize spacecraft thermal response space to the continuous solar exposure.

Midcourse corrections 1 and 2 will not normally be made unless the predicted Mission Control Center 3 velocity change is greater than 25 feet-per-second.

Lunar Orbit Insertion (LOI)

The first of two lunar orbit insertion burns will be made at 75:45:43 GET at an altitude of 89 nm above the Moon. LOI-1 will have a nominal retrograde velocity change of 2,974 fps and will insert Apollo 10 into a 60x170-nm elliptical lunar orbit. LOI-2 two orbits later at 80:10:45 GET will circularize the orbit to 60 nm. The burn will be 138.5 fps retrograde. Both LOI maneuvers will be with the SPS engine near pericynthion when the spacecraft is behind the Moon and out of contact with MSFN stations.

Lunar Parking Orbit (LPO) and LM-Active Rendezvous

Apollo 10 will remain in lunar orbit about 61.5 hours, and in addition to the LM descent to eight nautical miles above the lunar surface and subsequent rendezvous with the CSM, extensive lunar landmark tracking tasks will be performed by the crew.

Following a rest period after the lunar orbit circularization, the LM will be manned by the command and lunar module pilot and preparations begun for undocking at 98:10 GET. Some 25 minutes of station keeping and CSM inspection of the LM will be followed by a 2.5 fps radially downward SM RCS maneuver, placing the LM and CSM in equiperiod orbits with a maximum separation of two miles (minifootball). At the midpoint of the minifootball, the LM descent propulsion system (DPS) will be fired retrograde 71 fps at 99:34 GET for the descent orbit insertion (DOI) to lower LM pericynthion to eight miles. The DPS engine will be fired at 10 per cent throttle setting for 15 seconds and at 40 per cent for 13 seconds.

LUNAR ORBIT INSERTION

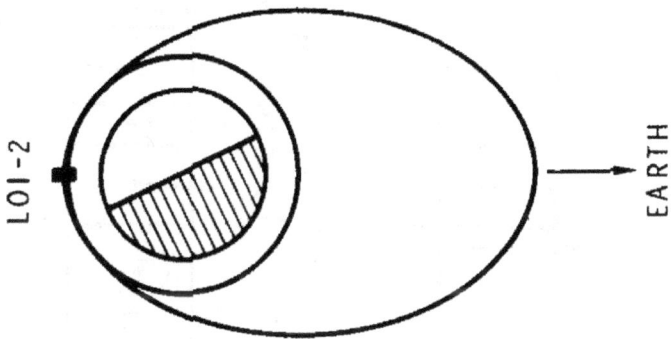

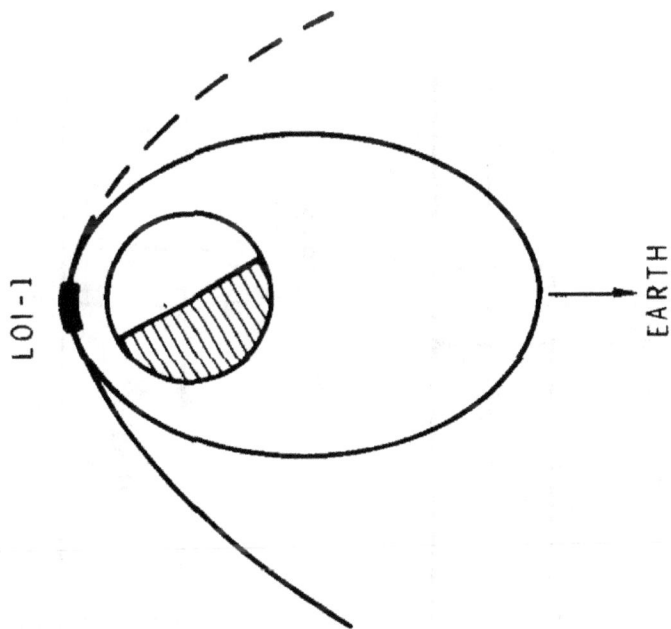

LOI-1

LOI-2

EARTH

EARTH

LUNAR ORBIT ACTIVITIES

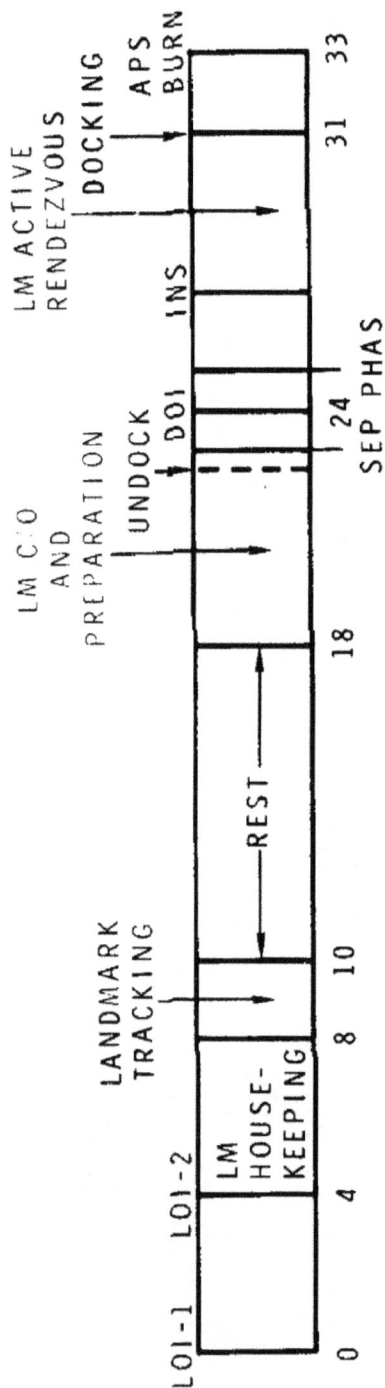

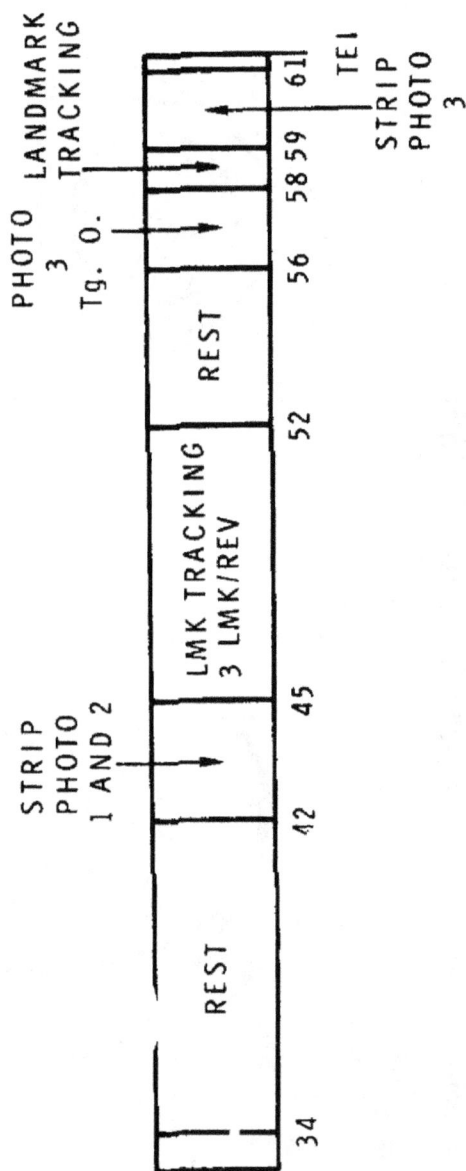

LUNAR ORBIT INSERTION

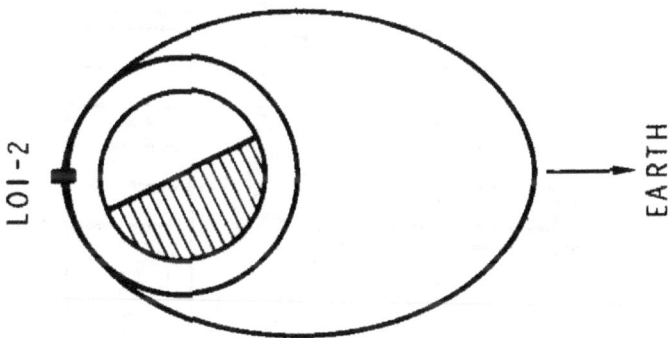

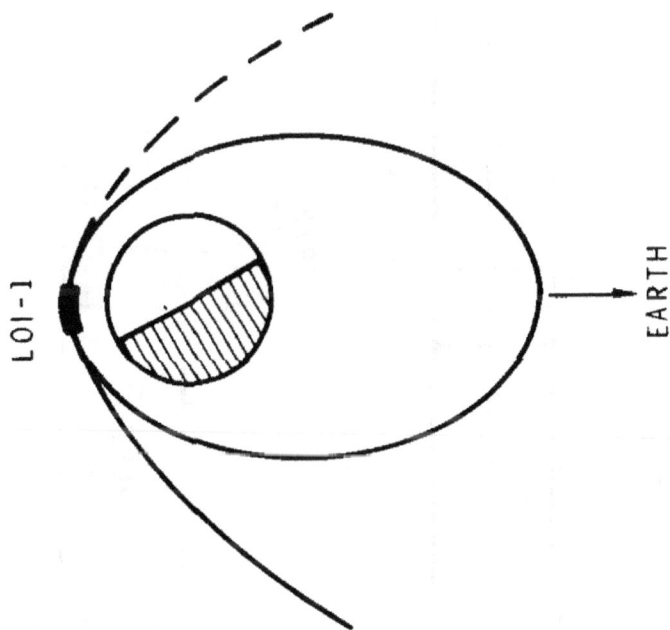

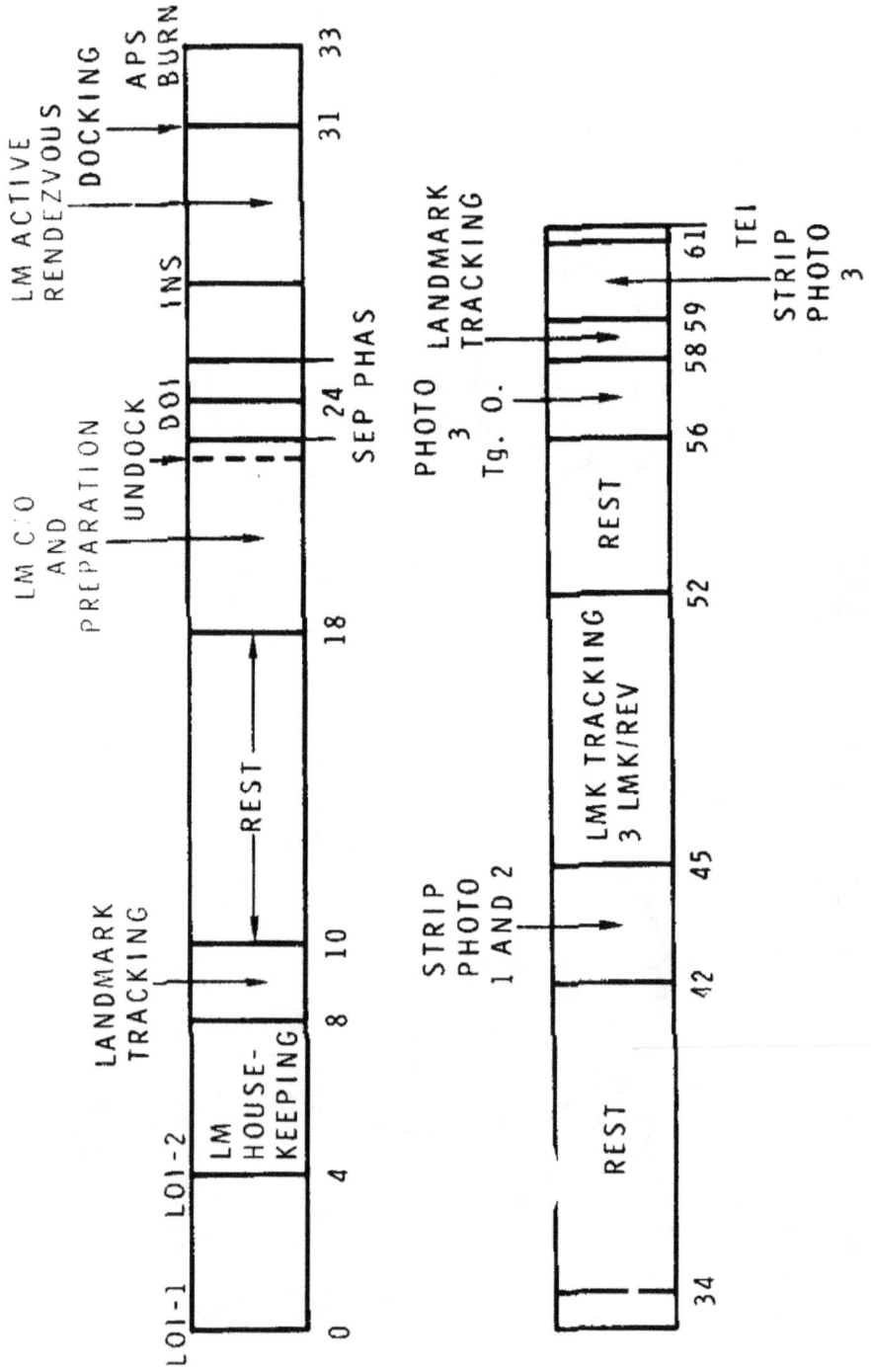

LUNAR ORBIT ACTIVITIES

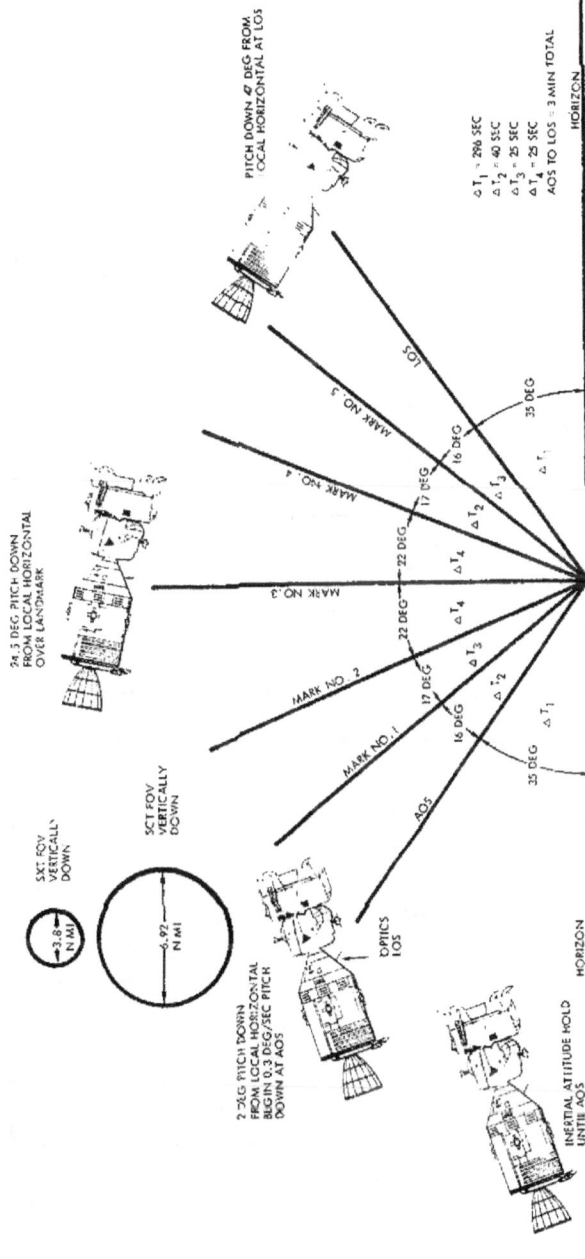

CSM/LM TYPICAL LANDMARK TRACKING PROFILE

APOLLO 10 RENDEZVOUS SEQUENCE

MANEUVER	G.E.T.	ΔV, FPS	ENGINE
SEPARATION	98:35:23	2.5	SM RCS
DOI	99:33:59	71.0	DPS
PHASING	100:46:21	195.0	DPS
INSERTION	102:43:18	207.0	APS
CSI	103:33:46	50.5	LM RCS
CDH	104:31:42	3.4	LM RCS
TPI	105:09:00	24.8	LM RCS
BRAKING	~105:55:00	~60.0	LM RCS
DOCKING	~106:20:00	~5.0	SM RCS

COMPARISON OF F AND G LM OPERATIONS PHASE

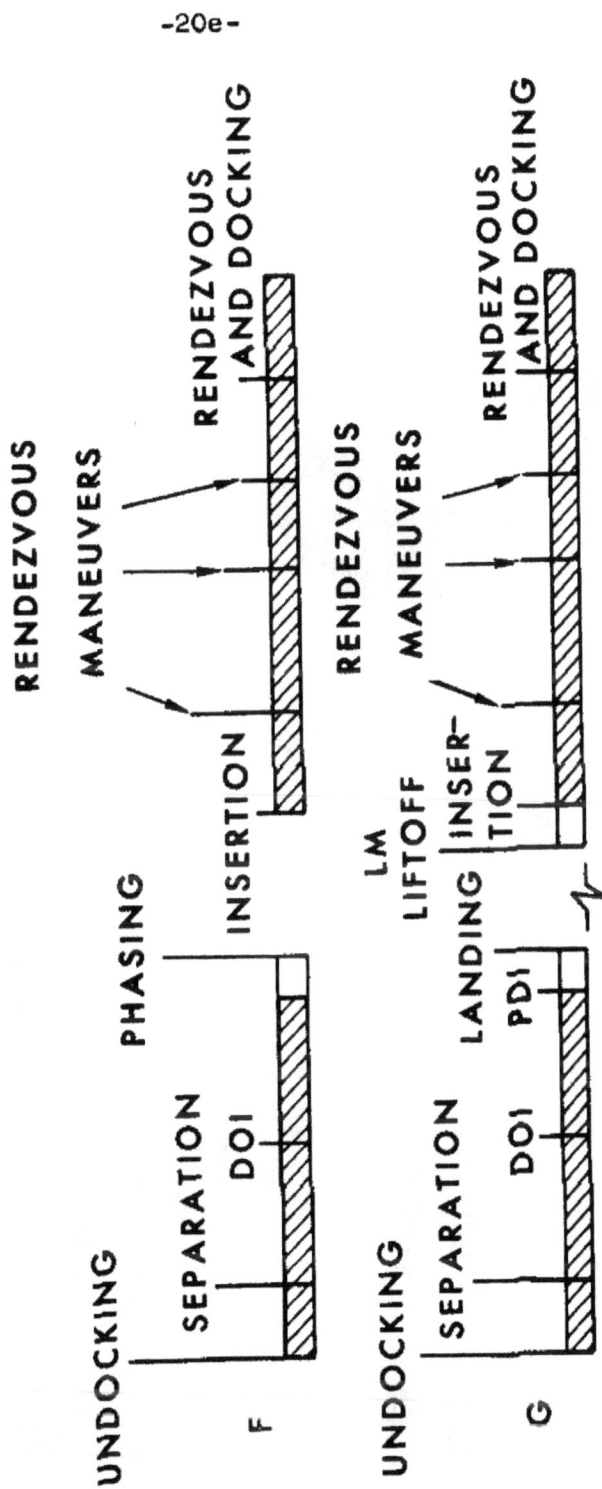

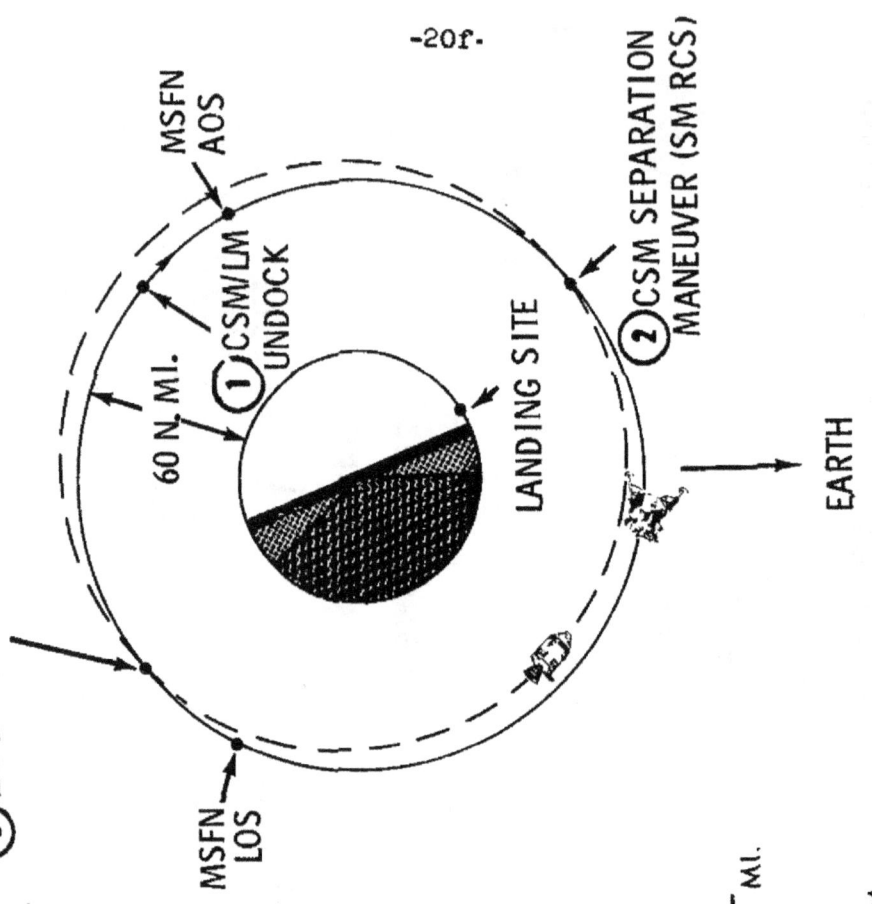

3 LM (DOI) DESCENT ORBIT INSERTION MANEUVER

MSFN LOS

MSFN AOS

1 CSM/LM UNDOCK

60 N. MI.

LANDING SITE

2 CSM SEPARATION MANEUVER (SM RCS)

EARTH

SURFACE DARKNESS

SC DARKNESS

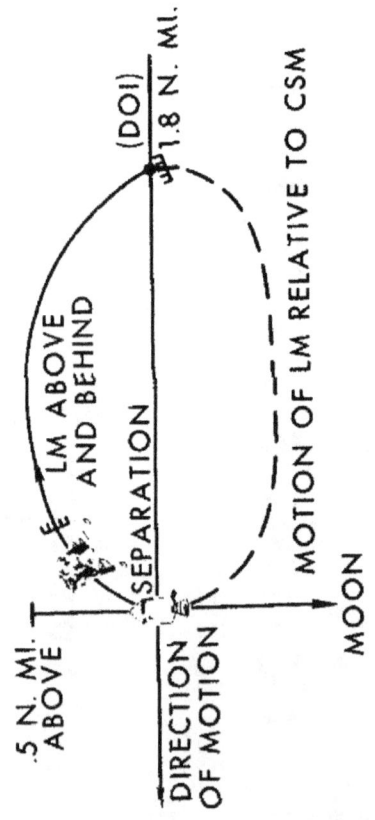

(DOI)

1.8 N. MI.

LM ABOVE AND BEHIND

.5 N. MI. ABOVE

SEPARATION

DIRECTION OF MOTION

MOON

MOTION OF LM RELATIVE TO CSM

CSM/LM SEPARATION MANEUVER

LUNAR MODULE DESCENT ORBIT INSERTION

-20g-

3 LM DESCENT ORBIT INSERTION
(DOI) MANUEVER,
RETROGRADE,
DPS –THROTTLED
TO 40%

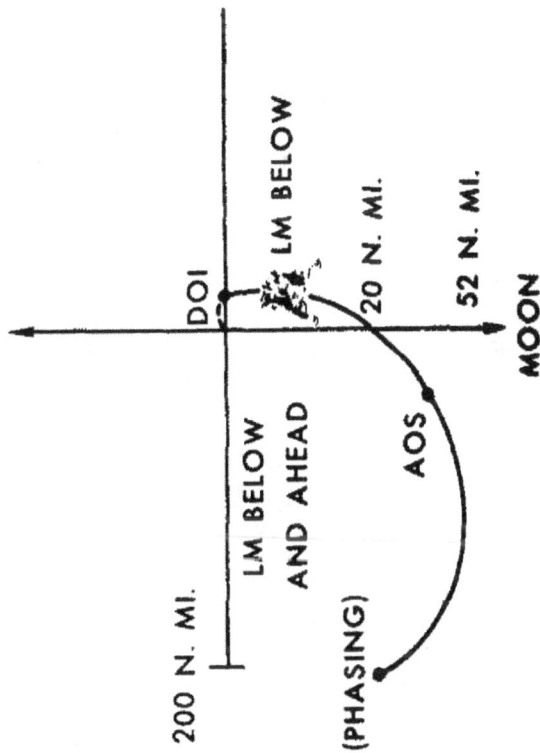

MSFN
AOS

MSFN
LOS

4 LM PHASING
MANEUVER

LANDING SITE

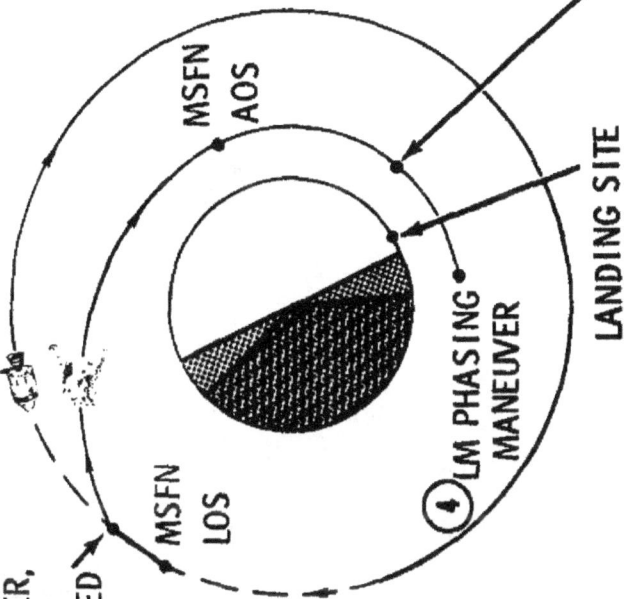

50,000 FT. ABOVE
LANDING SITE RADIUS

200 N. MI.

DOI

LM BELOW

LM BELOW
AND AHEAD

20 N. MI.

52 N. MI.

AOS

(PHASING)

MOON

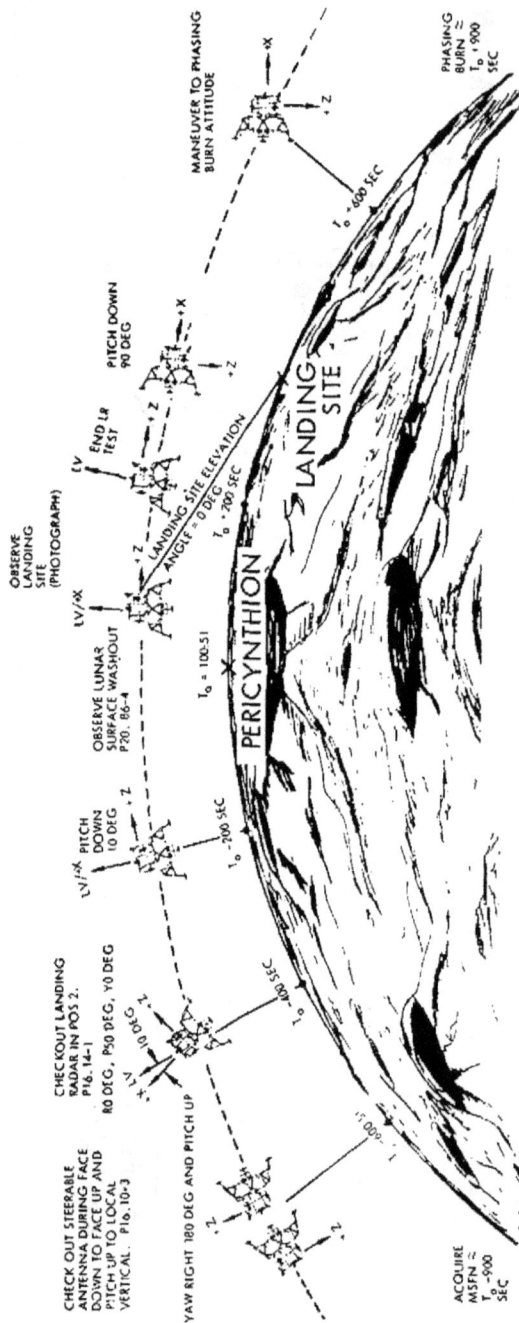

NEAR LUNAR SURFACE ACTIVITY

ORBIT RATE (0.05 DEG/SEC PITCH DOWN) FROM -400 to 200 FROM PERICYNTHION

As the LM passes over Apollo landing Site 2, the LM landing radar will be tested in the altitude mode but not in descent rate. About 10 minutes after the pass over Site 2, the 195 fps DPS phasing burn at 100:46 GET will boost the LM into an 8x194-nm orbit to allow the CSM to overtake and pass the LM. The phasing burn is posigrade and the DPS engine is fired at 10 per cent throttle for 26 seconds and full throttle for 17 seconds. The phasing burn places the LM in a "dwell" orbit which allows the CSM to overtake and pass the LM so that at the second LM passes over Site 2, the LM will trail the CSM by 27 nm and will be in a proper position for the insertion maneuver simulating ascent from the lunar surface after a landing mission.

Prior to the 207-fps LM ascent engine retrograde insertion burn, the LM descent stage will be jettisoned and an evasive maneuver performed by the ascent stage to prevent recontact. The insertion burn will be made at 102:43 GET and will lower LM apocynthion to 44.9 nm so that the LM is 14.7 nm below and 148 nm behind the CSM at the time of the concentric sequence initiate (CSI) burn.

Following LM radar tracking of the CSM and onboard computation of the CSI maneuver, a 50.5 fps LM RCS posigrade burn will be made at a nominal time of 103:33 GET at apocynthion and will result in a 44.9x44.3-nm LM orbit. The LM RCS will draw from the LM ascent propulsion system (APS) propellant tanks through the interconnect valves.

A 3.4 fps radially downward LM RCS constant delta height (CDH) maneuver at 104:31 GET will place the LM on a coelliptic orbit 15 nm below that of the CSM and will set up conditions for the terminal phase initiate (TPI) burn 38 minutes later.

The TPI maneuver will be made when the CSM is at a 26.6-degree elevation angle above the LM's local horizontal following continuing radar tracking of the CSM and onboard computations for the maneuver. Nominally, the TPI burn will be a 24.6-fps LM RCS burn along the line of sight toward the CSM at 105:09 GET. Midcourse correction and braking maneuvers will place the LM and CSM in a rendezvous and station-keeping position, and docking should take place at 106:20 GET to complete a eight-and-a-half hour sequence of undocked activities.

After the commander and lunar module pilot have transferred into the CSM, the LM will be jettisoned and the CSM will maneuver 2 fps radially upward to move above and behind the LM at the time of the LM ascent propulsion system burn to propellant depletion at 108:39 GET.

LUNAR MODULE PHASING MANEUVER

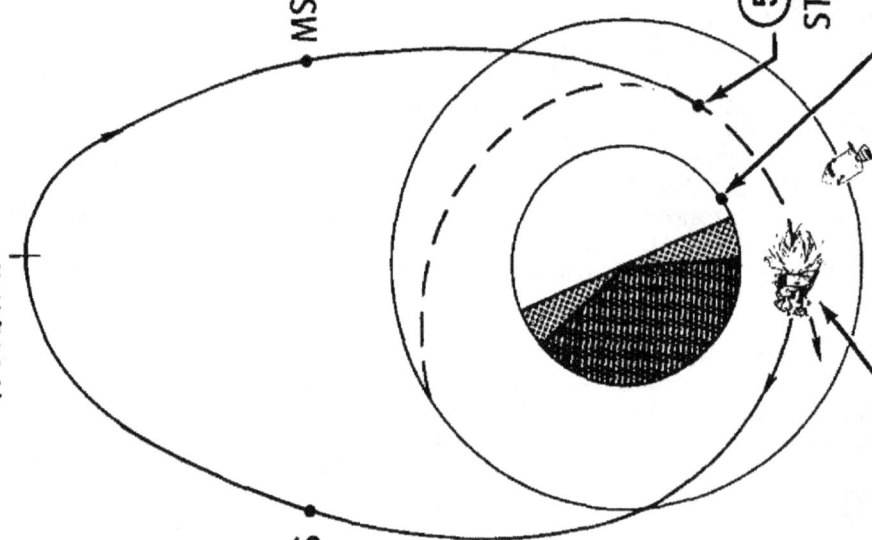

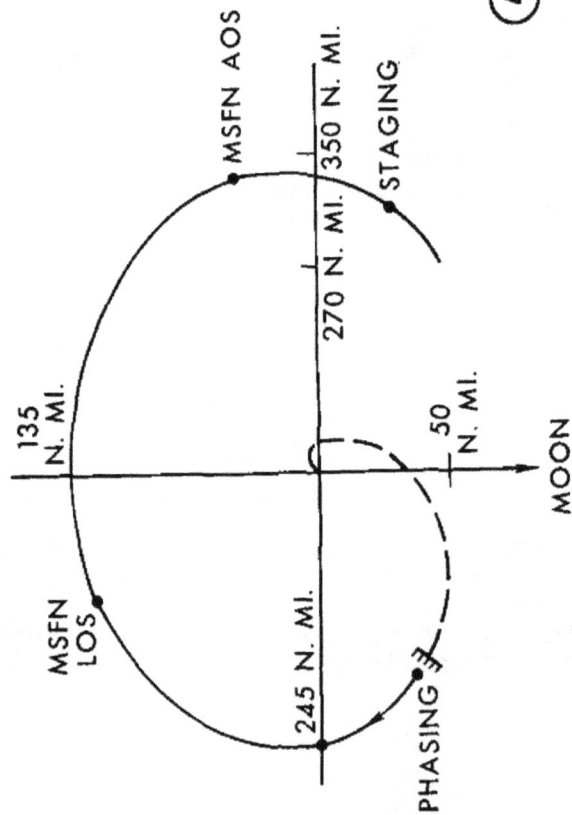

194 N. MI. APOCYNTHION

MSFN AOS

MSFN LOS

⑤ LM STAGING

LANDING SITE

④ LM PHASING MANEUVER DPS— FULL THROTTLE

MSFN AOS

350 N. MI.

STAGING

270 N. MI.

135 N. MI.

50 N. MI.

MOON

MSFN LOS

245 N. MI.

PHASING

LUNAR MODULE INSERTION MANEUVER

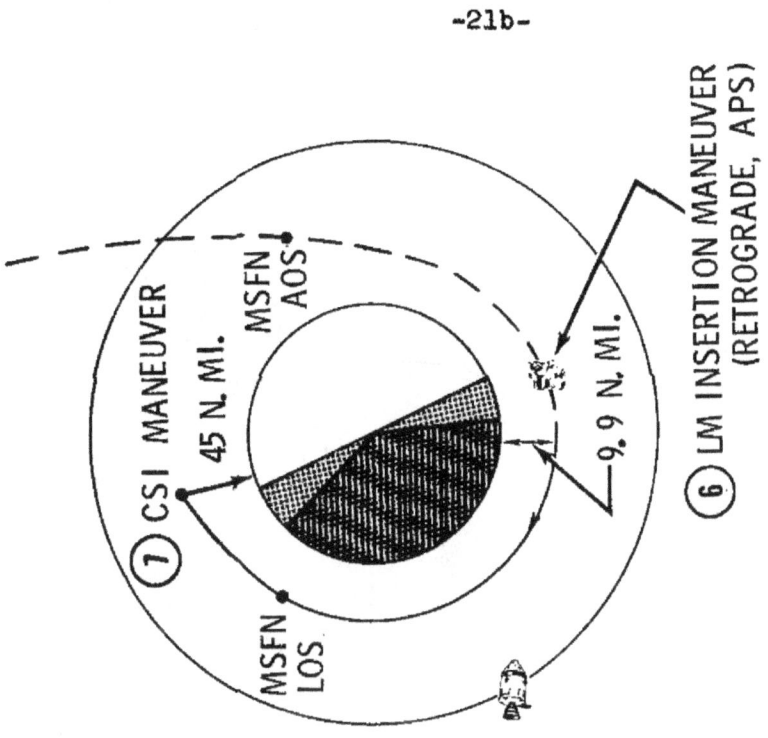

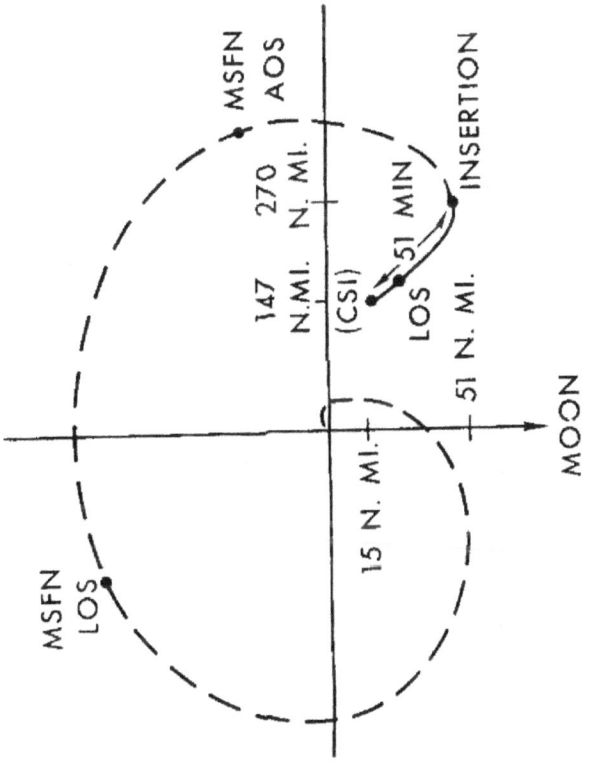

LUNAR MODULE
CONCENTRIC SEQUENCE INITIATION MANEUVER

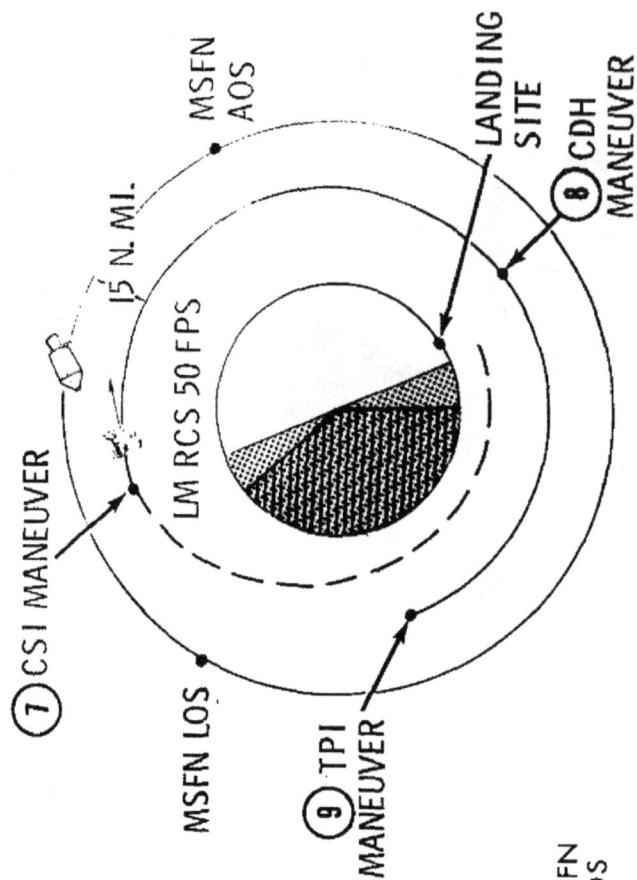

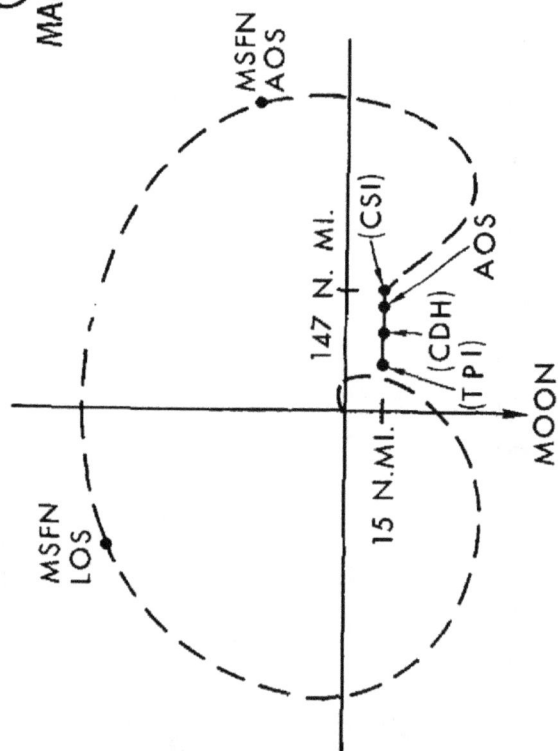

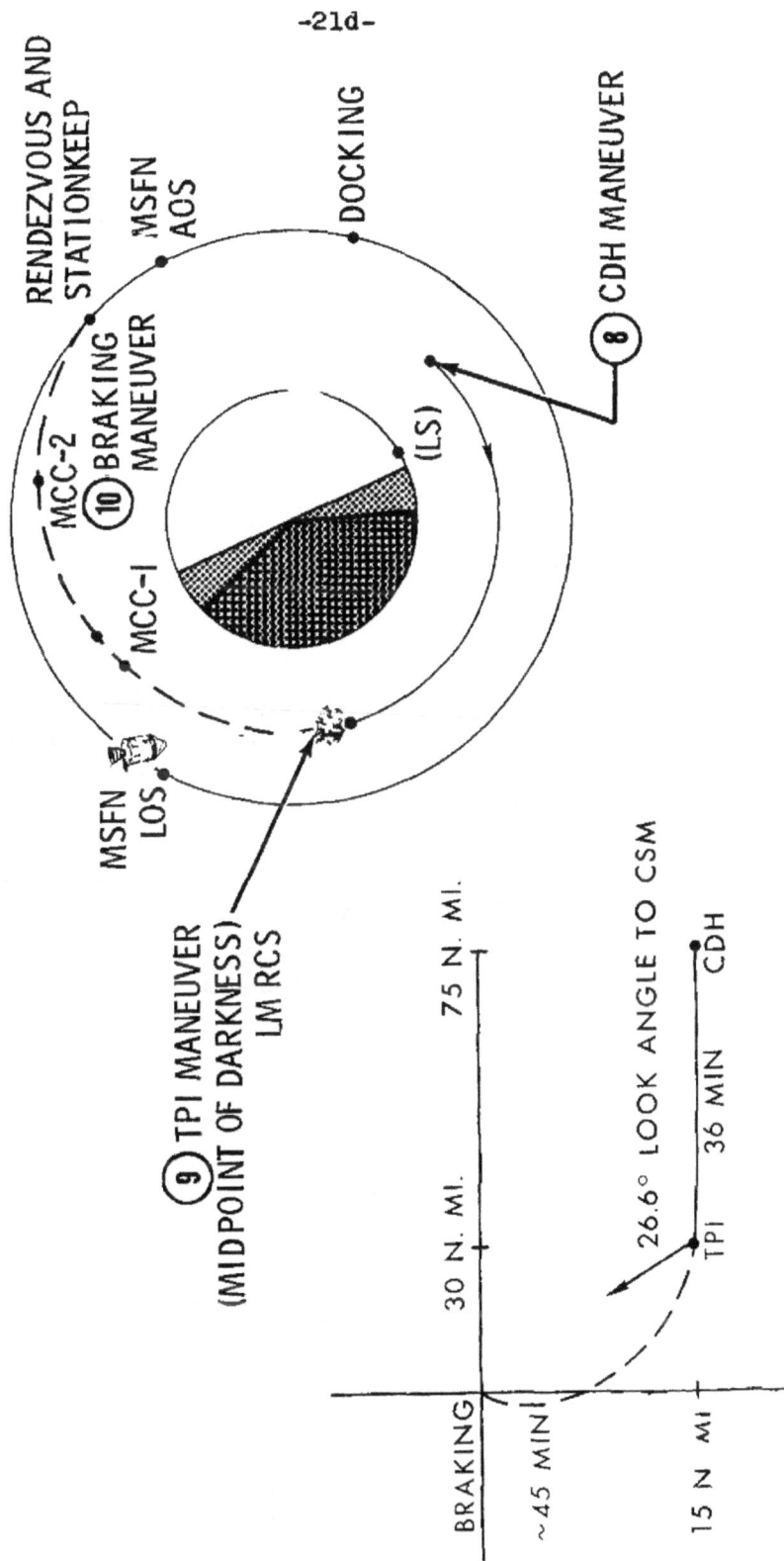

LUNAR MODULE CONSTANT

DIFFERENTIAL HEIGHT AND TERMINAL PHASE MANEUVERS

RENDEZVOUS AND STATIONKEEP

MSFN AOS

DOCKING

BRACKING MANEUVER ⑩

MCC-2

MCC-1

(LS)

⑧ CDH MANEUVER

MSFN LOS

⑨ TPI MANEUVER
(MIDPOINT OF DARKNESS)
LM RCS

75 N. MI.

30 N. MI.

26.6° LOOK ANGLE TO CSM

BRAKING

~45 MIN

15 N MI

TPI

36 MIN

CDH

The burn will be ground-commanded. An estimated 3,837-fps posigrade velocity will be imparted by the APS depletion burn near LM pericynthion to place the LM ascent stage in a helio-centric orbit.

An additional 29 hours will be spent in lunar orbit before transearth injection while the crew conducts lunar landmark tracking tasks and makes photographs of Apollo land-ing sites.

Transearth Injection (TEI)

The 54-hour return trip to Earth begins at 137:20 GET when the SPS engine is fired 3622.5 fps posigrade for the TEI burn. Like LOI-1 and LOI-2, the TEI burn will be made when the spacecraft is behind the Moon and out of touch with MSFN stations.

Transearth Coast

Three corridor-control transearth midcourse correction burns will be made if needed: MCC-5 at TEI +15 hours, MCC-6 at entry interface (EI=400,000 feet) -15 hours and at EI -3 hours.

Entry, Landing

Apollo 10 will encounter the Earth's atmosphere (400,000 feet) at 191:50 GET at a velocity of 36,310 fps and will land some 1,285 nm downrange from the entry-interface point using the spacecraft's lifting characteristics to reach the landing point. Touchdown will be at 192:05 GET at 15 degrees 7 minutes South latitude by 165 degrees West longitude.

EARTH ENTRY

- ENTRY RANGE CAPABILITY - 1200 TO 2500 N. MI.

- NOMINAL ENTRY RANGE - 1285 N. MI.

- SHORT RANGE SELECTED FOR NOMINAL MISSION BECAUSE:

  - RANGE FROM ENTRY TO LANDING CAN BE SAME FOR PRIMARY AND BACKUP CONTROL MODES

  - PRIMARY MODE EASIER TO MONITOR WITH SHORT RANGE

- WEATHER AVOIDANCE, WITHIN ONE DAY PRIOR TO ENTRY, IS ACHIEVED USING ENTRY RANGING CAPABILITY TO 2500 N. MI.

- UP TO ONE DAY PRIOR TO ENTRY USE PROPULSION SYSTEM TO CHANGE LANDING POINT

VELOCITY AT ENTRY INTERFACE

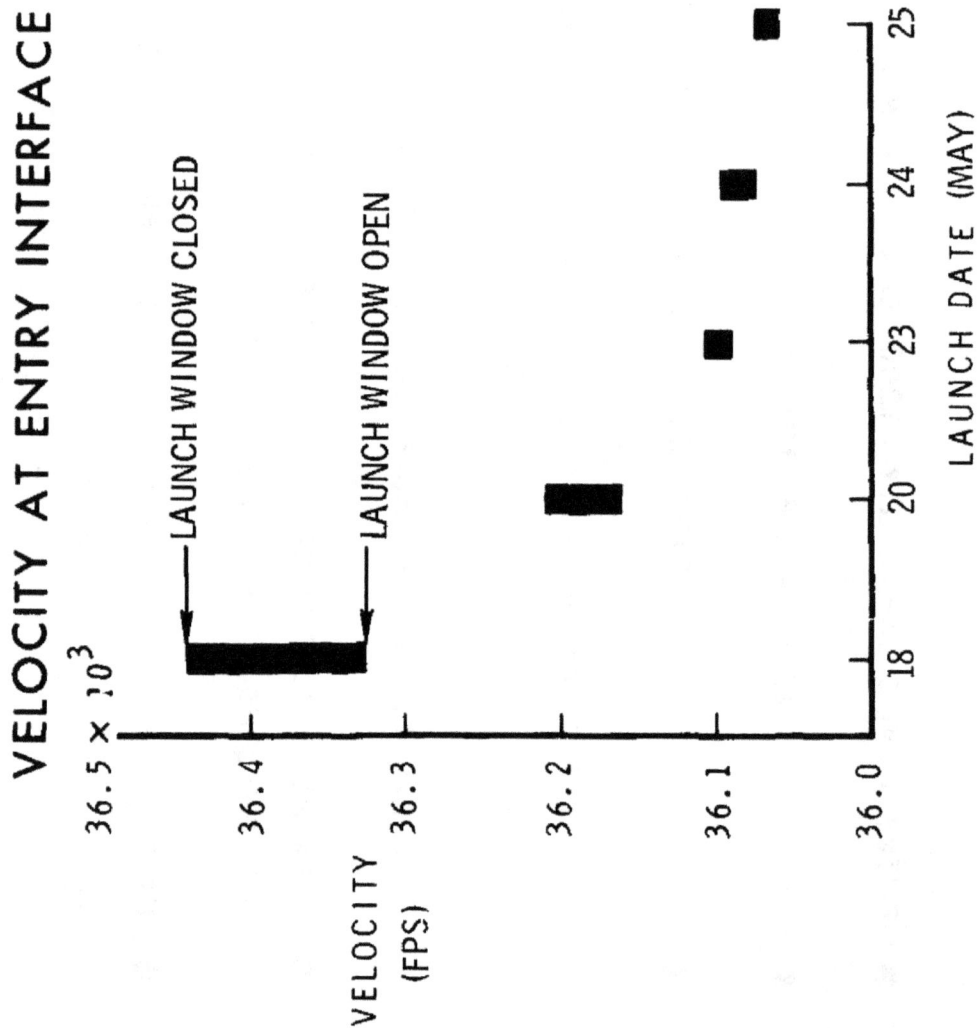

GEODETIC ALTITUDE VERSUS RANGE TO GO

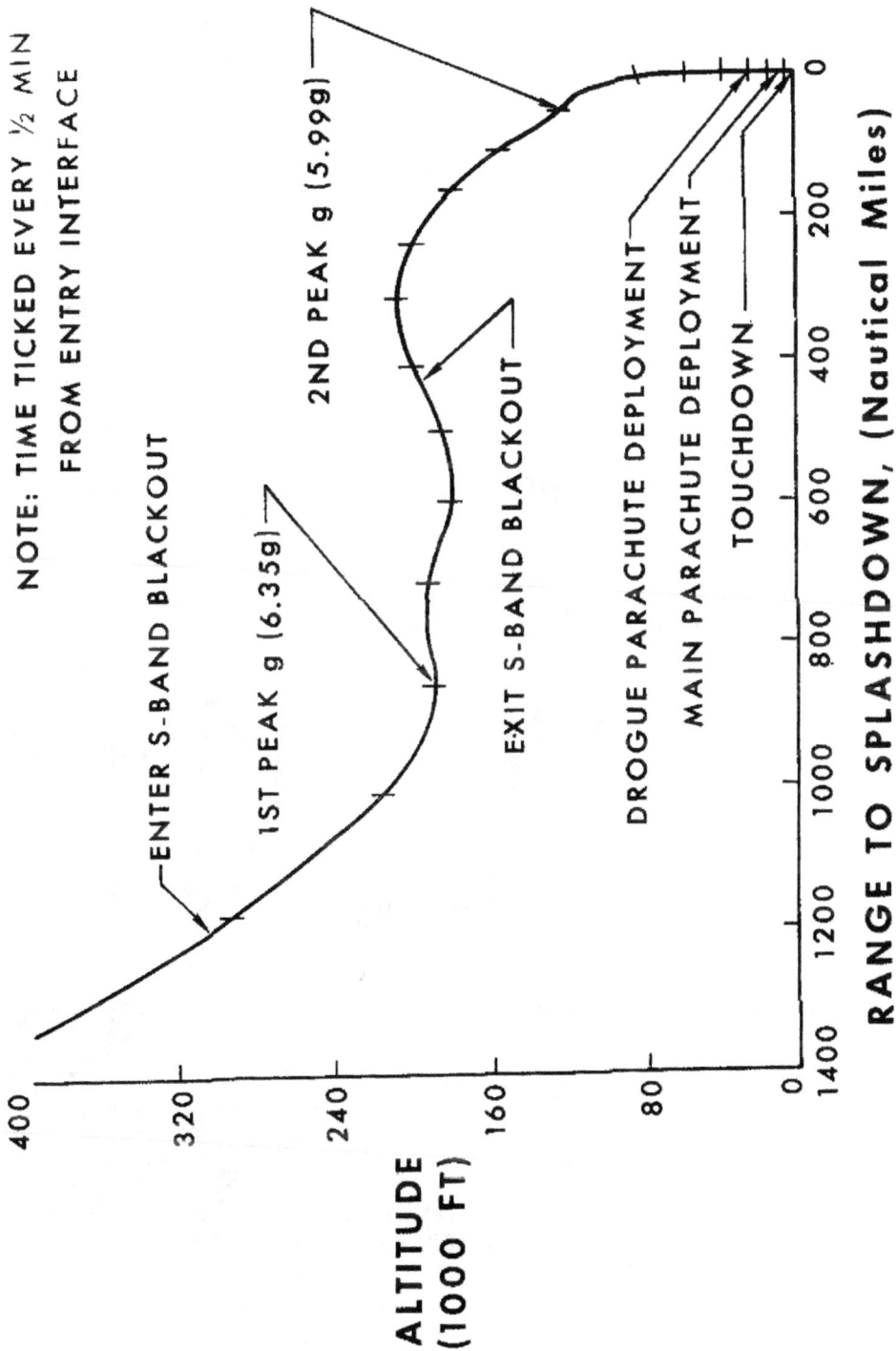

NOTE: TIME TICKED EVERY ½ MIN FROM ENTRY INTERFACE

ENTER S-BAND BLACKOUT

1ST PEAK g (6.35g)

2ND PEAK g (5.99g)

EXIT S-BAND BLACKOUT

DROGUE PARACHUTE DEPLOYMENT

MAIN PARACHUTE DEPLOYMENT

TOUCHDOWN

ALTITUDE (1000 FT)

400

320

240

160

80

0

RANGE TO SPLASHDOWN, (Nautical Miles)

1400   1200   1000   800   600   400   200   0

RANGE TO SPLASHDOWN, (Nautical Miles)

-more-

LIFT

DRAG

N MI (K)

400

300

ALT (FT)
(K)

200

100

0

2.0°

5.2°

DROGUE
CHUTES

PILOT CHUTES

DRAG
CHUTE

MAIN
CHUTES
(REEFED)

MAIN
CHUTES

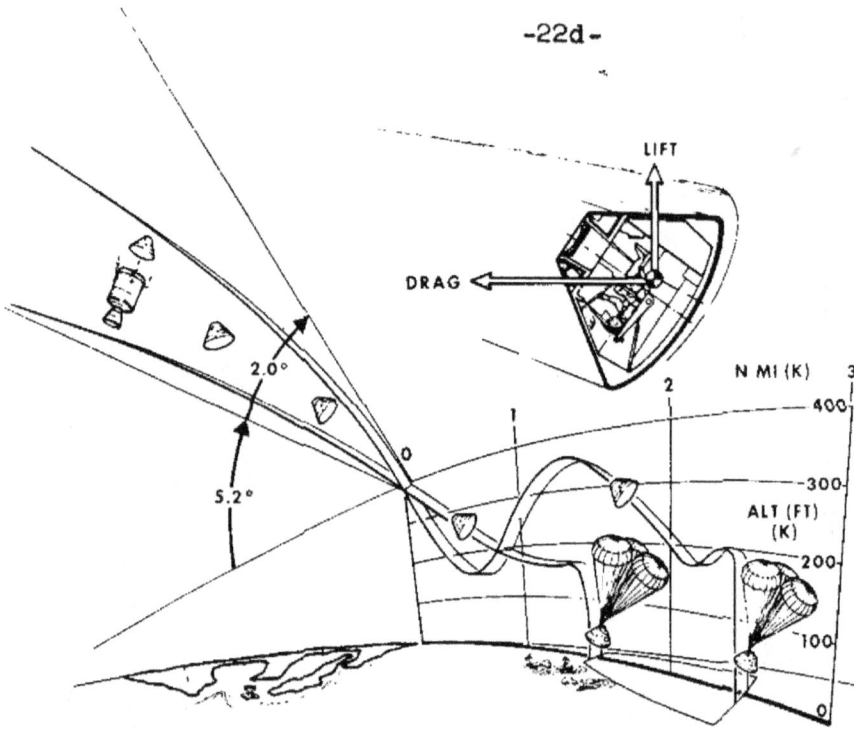

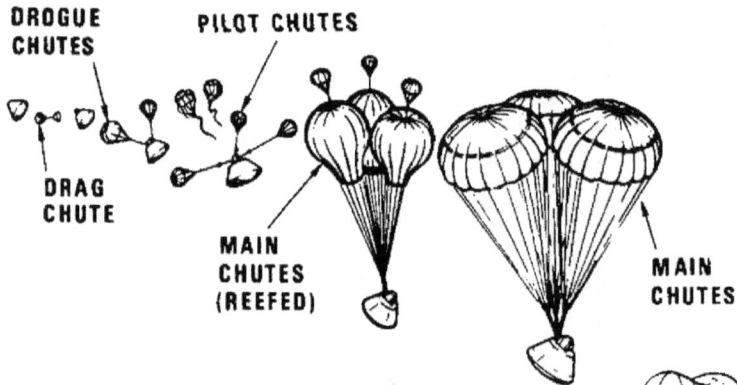

SPLASH DOWN VELOCITIES:
3 CHUTES - 31 FT/SEC
2 CHUTES - 36 FT/ SEC

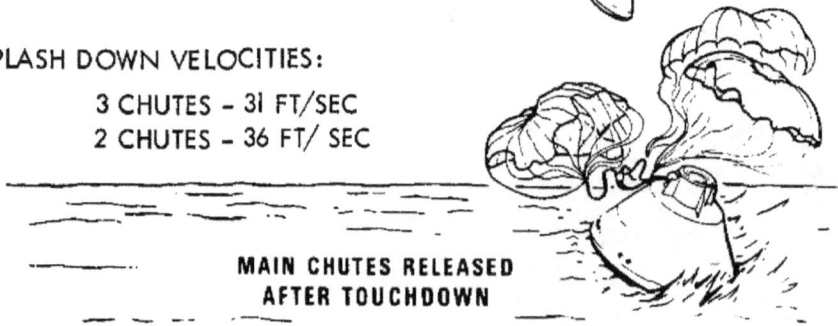

MAIN CHUTES RELEASED
AFTER TOUCHDOWN

EARTH RE-ENTRY AND LANDING

-more-

LM (SNOOPY)

- OPEN HATCH LMP INT TO LM
- LM STATUS CHECK, RECLINE AND STOW EQUIP
- IVT TO CSM
- LMP AND CDR IVT TO LM
- ACTIVATE EPS AND S-BAND OMNI
- ACTIVATE PGNCS AND RCS
- ERASABLE MEMORY DUMP
- ACTIVATE ASCENT BATTERY
- S-BAND STEERABLE CHECKS
- FPGA AND INITIALIZE AGS
- ACTIVATE AND INITIALIZE AGS
- RCS AND RR CHECKS
- AGS CALIBRATION AND ALIGN
- DPS PRESSURIZATION AND CHECKOUT
- 100% Q LANDING GEAR
- SUNDOCK
- IMU/AOT REALIGN-P52
- IMU AND VHF RANGING CHECKS
- RR TRACKING
- DOI BURN - DPS
- NEAR LUNAR SURFACE OBSERVATION AND PHOTOGRAPHY
- IMU/AOT REALIGN - P52, COAS CALIB
- PHASING BURN - DPS
- TAPS PRESS., PARALLEL ASCENT AND DESCENT BATS
- LM STAGING - RCS
- INSERTION BURN - APS
- IMU/AOT REALIGN
- ICSH BURN - RCS
- PLANE CHANGE - RCS
- ICSH BURN - RCS
- TPI BURN - RCS
- RENDEZVOUS MCC-1 AND MCC-2
- REENDEZVOUS
- DOCK
- S-BAND COMM TEST
- PREP LM AND APS BURN, IVT TO CSM
- JETTISON LM, APS BURN TO DEPL

CSM (CHARLIE BROWN)

- EARTH ORBIT AND SYSTEMS CHECKOUT
- TRANSPOSITION, DOCKING AND LM EXTRACTION
- ICSM/LM EVASIVE MANEUVER FROM S-IVB
- CISLUNAR NAVIGATION OPTICAL SIGHTING-STAR/EARTH HORIZON (5 SETS)
- MCC-1
- ESTABLISH PTC MODE
- MCC-2
- CISLUNAR NAVIGATION OPTICAL SIGHTING-STAR/EARTH HORIZON (5 SETS)
- PTC TEST
- PTC TEST
- MCC-3
- MCC-4
- IOI-1
- IOI-2
- IMU REALIGN
- OPEN HATCH INSPECT LATCHES
- TRANSFER EQUIP TO LM
- PSEUDO LANDMARK TRACKING
- COMM RELAY TEST
- CLOSE HATCH MAINTAIN LM PRESS
- PRE SLEEP CHECKS
- PHOTOGRAPHY, SLEEP COMM TEST
- IMU REALIGN
- IMU REALIGN P52
- INSTALL DROGUE PROBE AND HATCH
- LANDING SITE TRACKING
- SUNDOCK
- CSM SEPARATION MANEUVER
- INSPECT LM AND PHOTOGRAPHY
- IMU AND VHF RANGING C/O
- IMU REALIGN-P52
- EXT TRACKING/VHF RANGING
- LM PHASING BURN
- BACKUP LM INSERTION BURN
- BACKUP LM CSI BURN
- BACKUP LM CDH BURN
- BACKUP LM TPI BURN
- RENDEZVOUS - LM PHOTOGRAPHY
- DOCK
- PREPARE FOR LM JETTISON
- LM JETTISON
- LM JETTISON-RCS

CSM MANEUVER DATA

- T/L I-SIVB / 2:50:26 / BT: 5 MIN 22 SEC / ΔV: 10,089 FPS
- EVASIVE MANEUVER - SPS (CSM/LM FROM S-IVB) / 4:39:48 / BT: 2.8 SEC / ΔV: 19.7 FPS / NO ULLAGE
- MCC-1
- MCC-2 / 28:22 / BT: / ΔV:
- MCC-3 / 52:45
- MCC-4 / 70:45
- IOI-1 - SPS / 25:45:43 / BT: 5 MIN 2 SEC / ΔV: 2974.0 FPS / NO ULLAGE
- IOI-2 - SPS / 80:10:43 / BT: 16.4 SEC / ΔV: 138.5 FPS / ULLAGE: 2 JET 17 SEC
- CSM SEP-RCS / 96:03:36 / BT: 4.9 SEC / ΔV: 2.5 FPS / NO ULLAGE
- MCC-1 - SPS / 11:02:49 / BT: 8.1 SEC / ΔV: 5.7.0 FPS / NO ULLAGE
- LM JETTISON-RCS / 108:07:24 / BT: 5.6 SEC / ΔV: 2.0 FPS / NO ULLAGE

GET	0	4	8	12	16	20	24	28	32	36	40	44	48	52	56	60	64	68	72	76	80	84	88	92	96	100	104	108	112	116
LUNAR REVOLUTION NO.																				1	2	3	4	5	6	7	8	9	10	11	12	13	14	15	16	17	18	19	20	21	22
ALTITUDE NM x 10³	20	61	92	117	138	156	172	188	202										LUNAR ORBIT 209.6 - 215.8

—more—

RECOVERY OPERATIONS

The primary recovery line for Apollo 10 is in the mid-Pacific along the 175th West meridian of longitude above 15 degrees North latitude, and jogging to 165 degrees West longitude below the Equator. The helicopter carrier USS Princeton, Apollo 10 prime recovery vessel, will be stationed near the end-of-mission aiming point.

Splashdown for a full-duration lunar orbit mission launched on time May 18 will be at 5 degrees 8 minutes South by 165 degrees West at a ground elapsed time of 192 hours 5 minutes.

The latitude of splashdown depends upon the time of the transearth injection burn and the declination of the Moon at the time of the burn. A spacecraft returning from a lunar flight will enter Earth's atmosphere and splash down at a point on Earth directly opposite the Moon.

This point, called the antipode, is a projection of a line from the center of the Moon through the center of the Earth to the surface opposite the Moon. The mid-Pacific recovery line rotates through the antipode once each 24 hours, and the trans-Earth injection burn will be targeted for splashdown along the primary recovery line.

Other planned recovery lines for a deep-space mission are the East Pacific line extending roughly parallel to the coastlines of North and South America; the Atlantic Ocean line running along the 30th West meridian in the northern hemisphere and along the 25th West meridian in the southern hemisphere; the Indian Ocean line along the 65th East meridian; and the West Pacific line along the 150th East meridian in the northern hemisphere and jogging to the 170th East meridian in the southern hemisphere.

Secondary landing areas for a possible Earth orbital alternate mission have been established in two zones--one in the Pacific and one in the Atlantic.

Launch abort landing areas extend downrange 3,400 nautical miles from Kennedy Space Center, fanwise 50 nautical miles above and below the limits of the variable launch azimuth (72 degrees - 107 degrees). Ships on station in the launch abort area will be the destroyer USS Rich, the insertion tracking ship USNS Vanguard and the attack transport USS Chilton.

In addition to the primary recovery vessel steaming up and down the mid-Pacific recovery line and surface vessels on the Atlantic Ocean recovery line and in the launch abort area, 14 HC-130 aircraft will be on standby at seven staging bases around the Earth: Guam, Pago Pago, American Samoa; Hawaii, Bermuda; Lajes, Azores; Ascension Island; Mauritius and the Panama Canal Zone.

Apollo 10 recovery operations will be directed from the Recovery Operations Control Room in the Mission Control Center and will be supported by the Atlantic Recovery Control Center, Norfolk, Va., and the Pacific Recovery Control Center, Kunia, Hawaii.

The Apollo 10 crew will be flown from the primary recovery vessel to the Manned Spacecraft Center after recovery. The spacecraft will receive a preliminary examination, safing and power-down aboard the Princeton prior to offloading at Ford Island, Hawaii, where the spacecraft will undergo a more complete deactivation. It is anticipated that the spacecraft will be flown from Ford Island to Long Beach, Calif., within 72 hours, and then trucked to the North American Rockwell Space Division plant in Downey, Calif., for postflight analysis.

APOLLO 10 ALTERNATE MISSIONS

Five alternate mission plans have been prepared for the Apollo 10, each depending upon when in the mission time line it becomes necessary to switch to the alternate. Testing of the lunar module and a LM-active rendezvous in Earth orbit are preferred over a CSM-only flyby mission. When it is impossible to return to a low Earth orbit with rendezvous, a high-ellipse LM test is preferred over a low Earth orbit test.

Where possible, Apollo 10 alternate missions follow the lunar orbit mission time line and have a duration of about 10 days.

Apollo 10 alternate missions are summarized as follows:

Alternate 1:   Early shutdown of S-IVB during TLI with resulting apogee less than 25,000 nautical miles, or failure of S-IVB to insert spacecraft into Earth parking orbit and subsequent SPS contingency orbit insertion (COI), and in both cases no LM extraction possible. Alternate maneuvers would include:

* SPS phasing burn to obtain ground coverage of simulated lunar orbit insertion.

* Simulated LOI burn to a 100x400 nm Earth orbit.

* Midcourse corrections to modify orbit to 90x240 nm end-of-mission ellipse and to complete SPS lunar mission duty cycle during remainder of ten-day mission.

Alternate 2: S-IVB fails during TLI burn and resulting apogee is between 25,000 and 40,000 nautical miles; no LM extraction. Maneuver sequence would be:

* SPS phasing burn to obtain ground coverage of simulated lunar orbit insertion.

* Simulated LOI burn to a semi-synchronous Earth orbit.

* SPS phasing maneuver to place a later perigee over or opposite desired recovery zone.

* SPS maneuver to place CSM in semi-synchronous orbit with a 12-hour period.

* Deorbit directly from semi-synchronous orbit into Pacific recovery area (ten-day mission).

Alternate 3: No TLI burn or TLI apogee less than 4,000 nmi but LM successfully extracted.

* Simulated LOI burn to 100x400-nm orbit.

* Simulated descent orbit insertion (DOI) maneuver with LM.

* Simulated LM powered descent initiation (PDI) maneuver.

* Two SPS burns to circularize CSM orbit to 150 nm.

* LM-active rendezvous.

* Ground-commanded LM ascent propulsion system (APS) burn to depletion under abort guidance system (AGS) control, similar to APS depletion burn in Apollo 9.

* Additional SPS burns to place CSM in 90x240-nm end-of-mission ellipse and to complete SPS lunar mission duty cycle during remainder of ten-day mission.

Alternate 4:   Early S-IVB TLI cutoff with resulting apogee greater than 4,000 nm but less than 10,000 nm, and capability of SPS and LM descent propulsion system together to return CSM-LM to low Earth orbit without compromising CSM's ability to rescue LM.

* SPS phasing burn to obtain ground coverage of simulated lunar orbit insertion.

* First docked DPS burn out-of-plane simulates descent orbit insertion.

* Second docked DPS burn simulates power descent initiation.

* SPS simulated LOI burn.

* Phasing maneuver to obtain ground coverage of simulated powered descent initiation.

* SPS burns to circularize CSM orbit at 150 nm.

* LM-active rendezvous.

* Ground-commanded LM ascent propulsion system burn to depletion under abort guidance system (AGS) control, similar to APS depletion burn in Apollo 9.

* Additional SPS burns to place CSM in 90x240 nm end-of-mission ellipse and to complete SPS lunar mission duty cycle during remainder of ten-day mission.

Alternate 5:   SPS and DPS jointly cannot place CSM-LM in low Earth orbit without compromising ability of CSM to rescue LM in a rendezvous sequence, and SPS fuel quantity is too low for a CSM-LM circumlunar mission.

* SPS phasing burn to obtain ground coverage of simulated lunar orbit insertion.

* Simulated lunar orbit insertion into semisynchronous orbit.

* SPS phasing burn to obtain ground coverage of simulated power descent initiation.

* First docked DPS burn out of plane simulates descent orbit insertion.

* Second docked DPS burn simulates power descent initiation and is directed out of plane.

* SPS phasing burn to place a later perigee over or opposite desired recovery zone.

* SPS maneuver to place CSM-LM in semi-synchronous orbit with a 12-hour period.

* Ground-commanded LM ascent propulsion system burn to depletion under abort guidance system control; posigrade at apogee.

* Additional midcourse corrections along a lunar mission time line and direct entry from high ellipse.

ABORT MODES

The Apollo 10 mission can be aborted at any time during the launch phase or terminated during later phases after a successful insertion into Earth orbit.

Abort modes can be summarized as follows:

Launch phase --

Mode I - Launch escape (LES) tower propels command module away from launch vehicle. This mode is in effect from about T-45 minutes when LES is armed until LES jettison at 3:07 GET and command module landing point can range from the Launch Complex 39B area to 520 nm (600 sm, 964 km) downrange.

Mode II - Begins when LES is jettisoned and runs until the SPS can be used to insert the CSM into a safe Earth orbit (9:22 GET) or until landing points threaten the African coast. Mode II requires manual separation, entry orientation and full-lift entry with landing between 400 and 3,200 nm (461-3,560 sm, 741-5,931 km) downrange.

Mode III - Begins when full-lift landing point reaches 3,200 nm (3,560 sm, 5,931 km) and extends through Earth orbital insertion. The CSM would separate from the launch vehicle, and if necessary, an SPS retrograde burn would be made, and the command module would be flown half-lift to entry and landing at approximately 3,350 nm (3,852 sm, 6,197 km) downrange.

Mode IV and Apogee Kick - Begins after the point the SPS could be used to insert the CSM into an Earth parking orbit -- from about 9:22 GET. The SPS burn into orbit would be made two minutes after separation from the S-IVB and the mission would continue as an Earth orbit alternate. Mode IV is preferred over Mode III. A variation of Mode IV is the apogee kick in which the SPS would be ignited at first apogee to raise perigee for a safe orbit.

Deep Space Aborts

Translunar Injection Phase --

Aborts during the translunar injection phase are only a remote possibility, but if an abort became necessary during the TLI maneuver, an SPS retrograde burn could be made to produce spacecrafe entry. This mode of abort would be used only in the event of an extreme emergency that affected crew safety. The spacecraft landing point would vary with launch azimuth and length of the TLI burn. Another TLI abort situation would be used if a malfunction cropped up after injection. A retrograde SPS burn at about 90 minutes after TLI shutoff would allow targeting to land on the Atlantic Ocean recovery line.

Translunar Coast phase --

Aborts arising during the three-day translunar coast phase would be similar in nature to the 90-minute TLI abort. Aborts from deep space bring into the play the Moon's antipode (line projected from Moon's center through Earth's center to opposite face) and the effect of the Earth's rotation upon the geographical location of the antipode. Abort times would be selected for landing when the antipode crosses 165° West longitude. The antipode crosses the mid-Pacific recovery line once each 24 hours, and if a time-critical situation forces an abort earlier than the selected fixed abort times, landings would be targeted for the Atlantic Ocean, East Pacific, West Pacific or Indian Ocean recovery lines in that order of preference. When the spacecraft enters the Moon's sphere of influence, a circumlunar abort becomes faster than an attempt to return directly to Earth.

Lunar Orbit Insertion phase --

Early SPS shutdowns during the lunar orbit insertion burn (LOI) are covered by three modes in the Apollo 10 mission. All three modes would result in the CM landing at the Earth latitude of the Moon antipode at the time the abort was performed.

Mode I would be a LM DPS posigrade burn into an Earth-return trajectory about two hours (at next pericynthion) after an LOI shutdown during the first two minutes of the LOI burn.

Mode II, for SPS shutdown between two and three minutes after ignition, would use the LM DPS engine to adjust the orbit to a safe, non-lunar impact trajectory followed by a second DPS posigrade burn at next pericynthion targeted for the mid-Pacific recovery line.

Mode III, from three minutes after LOI ignition until normal cutoff, would allow the spacecraft to coast through one or two lunar orbits before doing a DPS posigrade burn at pericynthion targeted for the mid-Pacific recovery line.

Lunar Orbit Phase --

If during lunar parking orbit it became necessary to abort, the transearth injection (TEI) burn would be made early and would target spacecraft landing to the mid-Pacific recovery line.

Transearth Injection phase --

Early shutdown of the TEI burn between ignition and two minutes would cause a Mode III abort and a SPS posigrade TEI burn would be made at a later pericynthion. Cutoffs after two minutes TEI burn time would call for a Mode I abort---restart of SPS as soon as possible for Earth-return trajectory. Both modes produce mid-Pacific recovery line landings near the latitude of the antipode at the time of the TEI burn.

Transearth Coast phase --

Adjustments of the landing point are possible during the transearth coast through burns with the SPS or the service module RCS thrusters, but in general, these are covered in the discussion of transearth midcourse corrections. No abort burns will be made later than 24 hours prior to entry to avoid effects upon CM entry velocity and flight path angle.

APOLLO 10 GO/NO-GO DECISION POINTS

Like Apollo 8, Apollo 10 will be flown on a step-by-step commit point or go/no-go basis in which the decisions will be made prior to each major maneuver whether to continue the mission or to switch to one of the possible alternate missions. The go/no-go decisions will be made by the flight control teams in Mission Control Center.

Go/no-go decisions will be made prior to the following events:

* Launch phase go/no-go at 10 min. GET for orbit insertion

* Translunar injection

* Transposition, docking and LM extraction

* Each translunar midcourse correction burn

* Lunar orbit insertion burns Nos. 1 and 2

* Crew intravehicular transfer to LM

* CSM-LM undocking and separation

* Rendezvous sequence

* LM Ascent Propulsion system burn to depletion

* Transearth injection burn (no-go would delay TEI one or more revolutions to allow maneuver preparations to be completed.)

* Each transearth midcourse correction burn

ONBOARD TELEVISION

On Apollo 10, onboard video will originate from the CM; there will be no TV camera in the LM. Plans call for both black and white and color TV to be carried.

The black and white camera is a 4.5 pound RCA camera equipped with a 80-degree field of view wide angle and 100mm nine-degree field of view telephoto lens, attached to a 12-foot power/video cable. It produces a black-and-white 227 TV line signal scanned at 10 frames a second. Madrid, Goldstone and Honeysuckle Creek all will have equipment to make still photographs of the slow scan signal and to convert the signal to commercial TV format.

The color TV camera is a 12-pound Westinghouse camera with a zoom lens for close-up or wide angle use and a three-inch monitor which can be mounted on the camera or in the CM. It produces a standard 525-line, 30-frame-per-second signal in color by use of a rotating color wheel. The signal can be viewed in black and white. Only MSC, receiving the signal through Goldstone, will have equipment to colorize the signal.

Tentative planning is to use the color camera predominately, reverting to the black and white camera if there is difficulty with the color system but requiring at least one black and white transmission to Honeysuckle Creek. The following is a preliminary plan for TV passes based on a 12:49 May 18 launch:

GET	DATE/EDT	EVENT	
3:00 - 3:15	18 - 3:48p	Transposition & dock	Madrid
3:15 - 3:25	18 -	"	Goldstone
27:15 - 27:25	19 - 4:03p	Translunar coast	Goldstone
54:00 - 54:10	20 - 6:48p	Translunar coast	Goldstone
72:20 - 72:35	21 - 1:08p	Pre-LOI-1	Goldstone/Madrid
80:45 - 80:53	21 - 9:33p	Post LOI-2	Goldstone
98:15 - 98:20	22 - 3:01p	Post undock; formation	Goldstone
108:35 - 108:45	23 - 1:23a	APS Burn to Depletion	Goldstone
126:20 - 127:00	23 - 7:08p	Landmark Tracking	Goldstone
137:45 - 137:55	24 - 6:33a	Post-TEI	Honeysuckle*
152:35 - 152:45	24 - 9:23p	Transearth coast	Goldstone
186:50 - 187:00	26 - 7:38a	Transearth coast	Goldstone

*Transmission from RCA black and white camera. All others planned to be from color camera.

APOLLO 10 PHOTOGRAPHIC TASKS

Still and motion pictures will be made of most spacecraft
maneuvers as well as of the lunar surface and of crew activities
in the Apollo 10 cabin.

The transposition, docking and lunar module ejection
maneuver will be the first major event to be photographed. In
lunar orbit, the LM-active rendezvous sequence will be photographed from both the command and the lunar module.

During the period between the LM DPS phasing burn and the
APS insertion burn, the commander and lunar module pilot will
make still photos of the lunar ground track and of landing Site
2 from the eight-mile low point of the LM's flight path.

After rendezvous is complete and the LM APS depletion burn
has been photographed, the crew will make stereo strip still
photographs of the lunar surface and individual frames of targets
of opportunity. Using the navigation sextant's optics as a
camera lens system, lunar surface features and landmarks will be
recorded on motion picture film. Additionally, the camera-
through-sextant system will photograph star-horizon and star-land-
mark combinations as they are superimposed in visual navigation
sightings.

The Apollo 10 photography plan calls for motion pictures
of crew activities such as intravehicular transfer through the
CSM-LM docking tunnel and of other crew activities such as
pressure suit donning.

Long-distance Earth and lunar terrain photographs will be
shot with the 70mm still cameras.

Camera equipment carried on Apollo 10 consists of two 70mm
Hasselblad still cameras, each fitted with 80mm f/2.8 to f/22
Zeiss Planar lenses, a 250mm telephoto lens stowed aboard the
command module, and associated equipment such as filters, ring-
sight, spotmeter and an intervalometer for stereo strip photography.
One Hasselblad will be stowed in the LM and returned to the CSM
after rendezvous. Hasselblad shutter speeds range from one second
to 1/500 sec.

For motion pictures, two Maurer data acquisition cameras
(one in the CSM, one in the LM) with variable frame speed
selection will be used.. Motion picture camera accessories
include bayonet-mount lenses of 75, 18, and 5mm focal length,
a right-angle mirror, a command module boresight bracket, a
power cable, and an adapter for shooting through the sextant.

Apollo 10 film stowage includes six 70mm Hasselblad
magazines---two exterior color reversal and four fine-grain
black and white; and 12 140-foot 16mm magazines of motion
picture film---eight exterior color and four interior color---
for a total 1680 feet.

LUNAR DESCRIPTION

Terrain - Mountainous and crater-pitted, the former rising thousands of feet and the latter ranging from a few inches to 180 miles in diameter. The craters are thought to be formed by the impact of meteorites. The surface is covered with a layer of fine-grained material resembling silt or sand, as well as small rocks and boulders.

Environment - No air, no wind, and no moisture. The temperature ranges from 243 degrees in the two-week lunar day to 279 degrees below zero in the two-week lunar night. Gravity is one-sixth that of Earth. Micrometeoroids pelt the Moon (there is no atmosphere to burn them up). Radiation might present a problem during periods of unusual solar activity.

Dark Side - The dark or hidden side of the Moon no longer is a complete mystery. It was first photographed by a Russian craft and since then has been photographed many times, particularly by NASA's Lunar Orbiter spacecraft and Apollo 8.

Origin - There is still no agreement among scientists on the origin of the Moon. The three theories: (1) the Moon once was part of Earth and split off into its own orbit, (2) it evolved as a separate body at the same time as Earth, and (3) it formed elsewhere in space and wandered until it was captured by Earth's gravitational field.

Physical Facts

Diameter	2,160 miles (about ¼ that of Earth)
Circumference	6,790 miles (about ¼ that of Earth)
Distance from Earth	238,857 miles (mean; 221,463 minimum to 252,710 maximum)
Surface temperature	+243°F (Sun at zenith) -279°F (night)
Surface gravity	1/6 that of Earth
Mass	1/100th that of Earth
Volume	1/50th that of Earth
Lunar day and night	14 Earth days each
Mean velocity in orbit	2,287 miles per hour
Escape velocity	1.48 miles per second
Month (period of rotation around Earth)	27 days, 7 hours, 43 minutes

Apollo Lunar Landing Sites

Possible landing sites for the Apollo lunar module have been under study by NASA's Apollo Site Selection Board for more than two years. Thirty sites originally were considered. These have been narrowed down to four for the first lunar landing. (Site 1 currently not considered for first landing.)

Selection of the final five sites was based on high resolution photographs by Lunar Orbiter spacecraft, plus close-up photos and surface data provided by the Surveyor spacecraft which soft-landed on the Moon.

The original sites are located on the visible side of the Moon within 45 degrees east and west of the Moon's center and 5 degrees north and south of its equator.

The final site choices were based on these factors:

*Smoothness (relatively few craters and boulders)

*Approach (no large hills, high cliffs, or deep craters that could cause incorrect altitude signals to the lunar module landing radar)

*Propellant requirements (selected sites require the least expenditure of spacecraft propellants)

*Recycle (selected sites allow effective launch preparation recycling if the Apollo Saturn V countdown is delayed)

*Free return (sites are within reach of the spacecraft luanched on a free return translunar trajectory)

*Slope (there is little slope -- less than 2 degrees in the approach path and landing area)

The Five Landing Sites Finally Selected Are:

Designations	Center Coordinates

Site 1

latitude 2° 37' 54" North
longitude 34° 01' 31" East

Site 1 is located on the east central part of the Moon in southeastern Mare Tranquillitatis. The site is approximately 62 miles (100 kilometers) east of the rim of Crater Maskelyne.

Site 2

latitude 0° 43' 56" North
longitude 23° 38' 51" East

Site 2 is located on the east central part of the Moon in southwestern Mar Tranquillitatis. The site is approximately 62 miles (100 kilometers) east of the rim of Crater Sabine and approximately 118 miles (190 kilometers) southwest of the Crater Maskelyne.

Site 3

latitude 0° 22' 27" North
longitude 1° 20' 42" West

Site 3 is located near the center of the visible face of the Moon in the southwestern part of Sinus Medii. The site is approximately 25 miles (40 kilometers) west of the center of the face and 21 miles (50 kilometers) southwest of the Crater Bruce.

Site 4

latitude 3° 38' 34" South
longitude 36° 41' 53" West

Site 4 is located on the west central part of the Moon in southeastern Oceanus Procellarum. The site is approximately 149 miles (240 kilometers) south of the rim of Crater Encke and 136 miles (220 kilometers) east of the rim of Crater Flamsteed.

Site 5

latitude 1° 46' 19" North
longitude 41° 56' 20" West

Site 5 is located on the west central part of the visible face in southeastern Oceanus Procellarum. The site is approximately 130 miles (210 kilometers) southwest of the rim of Crater Kepler and 118 miles (190 kilometers) north northeast of the rim of Crater Flamsteed.

APOLLO LUNAR LANDING SITES

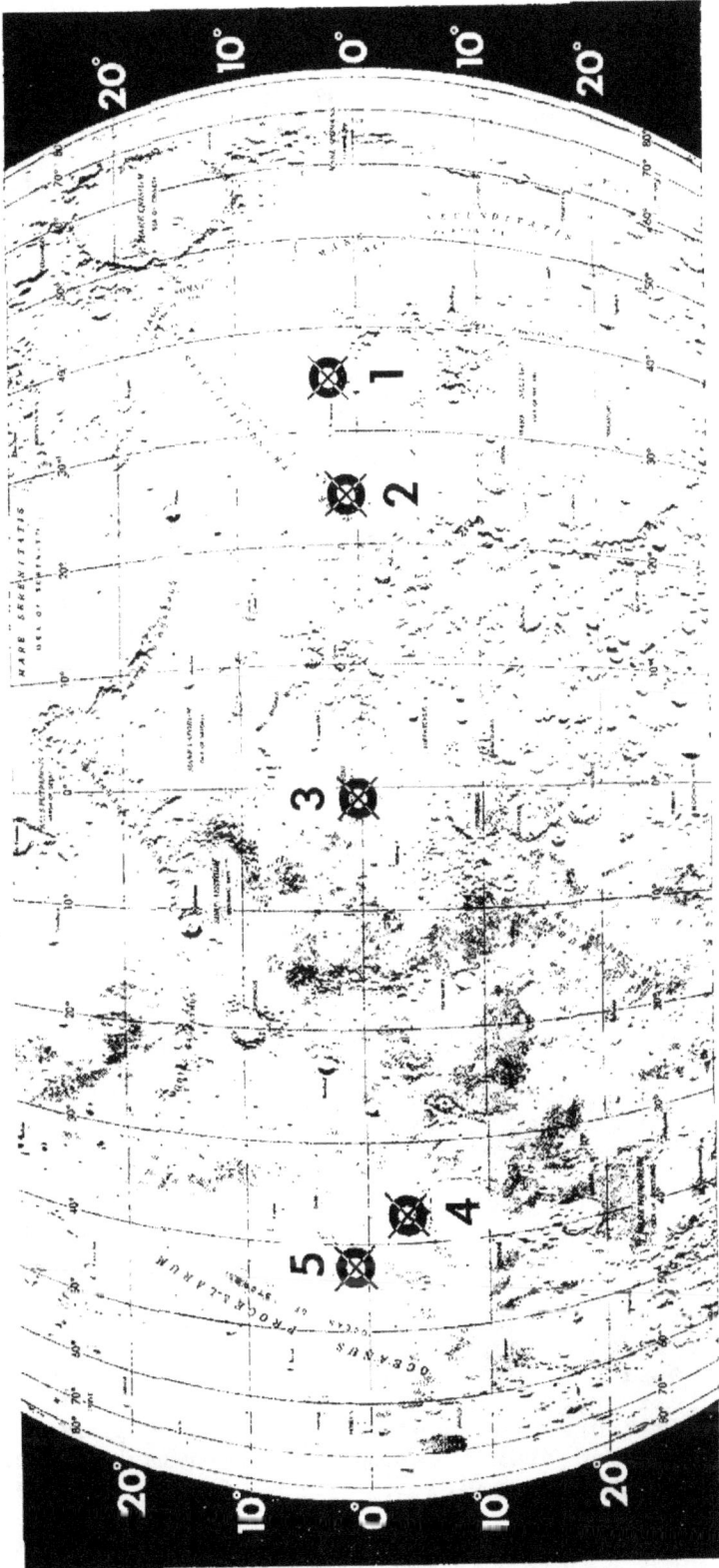

COMMAND AND SERVICE MOUDLE STRUCTURE, SYSTEMS

The Apollo spacecraft for the Apollo 10 mission is comprised of Command Module 106, Service Module 106, Lunar Module 4, a spacecraft-lunar module adapter (SLA) and a launch escape system. The SLA serves as a mating structure between the instrument unit atop the S-IVB stage of the Saturn V launch vehicle and as a housing for the lunar module.

Launch Escape System (LES) -- Propels command module to safety in an aborted launch. It is made up of an open-frame tower structure, mounted to the command module by four frangible bolts, and three solid-propellant rocket motors: a 147,000 pound-thrust launch escape system motor, a 2,400-pound-thrust pitch control motor, and a 31,500-pound-thrust tower jettison motor. Two canard vanes near the top deploy to turn the command module aerodynamically to an attitude with the heat-shield forward. Attached to the base of the launch escape tower is a boost protective cover composed of glass, cloth, and honeycomb, that protects the command module from rocket exhaust gases from the main and the jettison motors. The system is 33 feet tall, four feet in diameter at the base, and weighs 8,848 pounds.

Command Module (CM) Structure -- The basic structure of the command module is a pressure vessel encased in heat shields, cone-shaped 11 feet 5 inches high, base diameter of 12 feet 10 inches, and launch weight 12,277 pounds.

The command module consists of the forward compartment which contains two reaction control engines and components of the Earth landing system; the crew compartment or inner pressure vessel containing crew accomodations, controls and displays, and spacecraft systems; and the aft compartment housing ten reaction control engines and propellant tankage. The crew compartment contains 210 cubic feet of habitable volume.

Heat-shields around the three compartments are made of brazed stainless steel honeycomb with an outer layer of phenolic epoxy resin as an ablative material. Shield thickness, varying according to heat loads, ranges from 0.7 inch at the apex to 2.7 inches at the aft end.

The spacecraft inner structure is of sheet-aluminum honeycomb bonded sandwhich ranging in thickness from 0.25 inch thick at forward access tunnel to 1.5 inches thick at base.

CSM 106 and LM-4 are equipped with the probe-and-drogue docking hardware. The probe assembly is a folding coupling and impact attentuating device mounted on the CM tunnel that mates with a conical drogue mounted on the LM docking tunnel. After the docking latches are dogged down following a docking maneuver, both the probe and drogue assemblies are removed from the vehicle tunnels and stowed to allow free crew transfer between the CSM and LM.

APOLLO SPACECRAFT

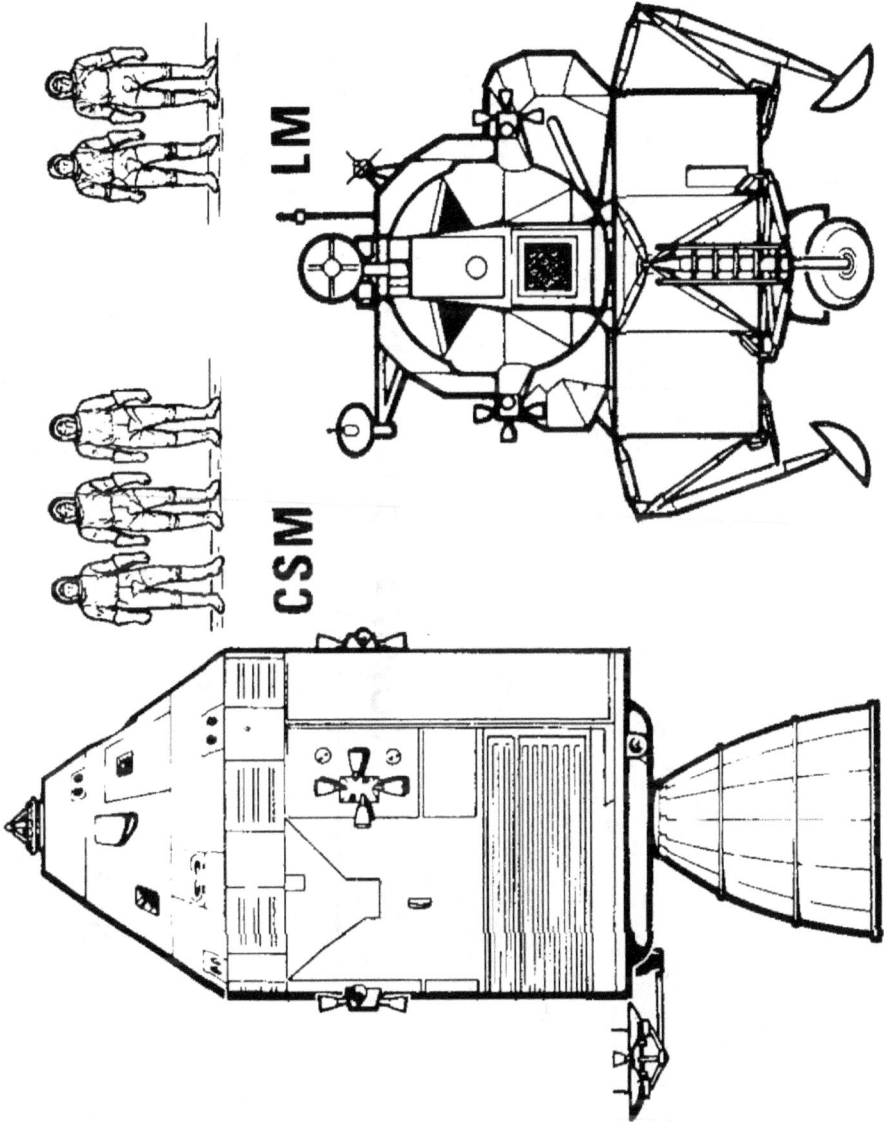

LM

CSM

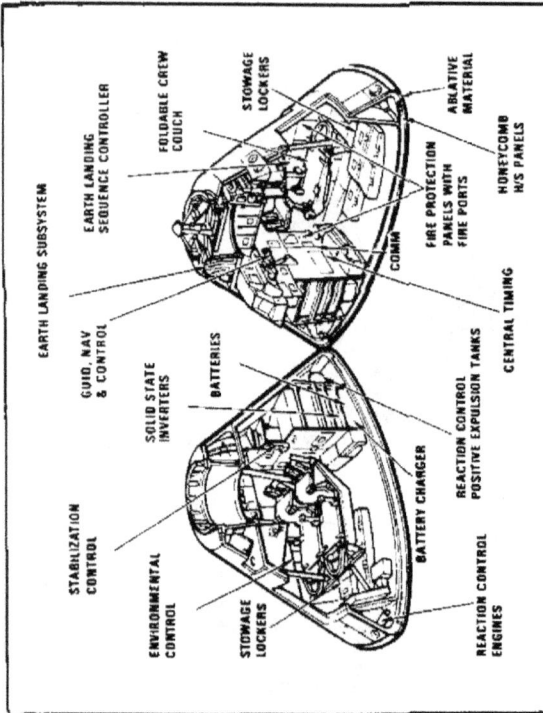

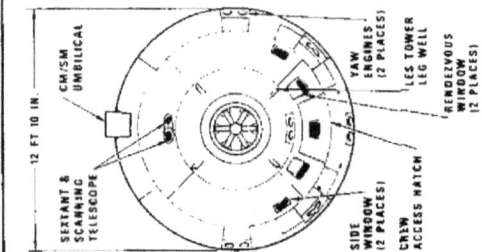

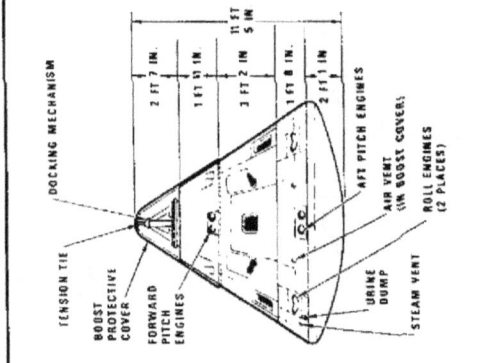

COMMAND MODULE

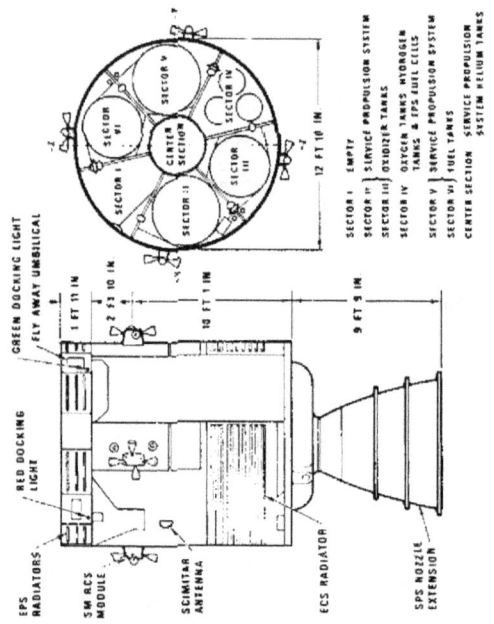

SERVICE MODULE

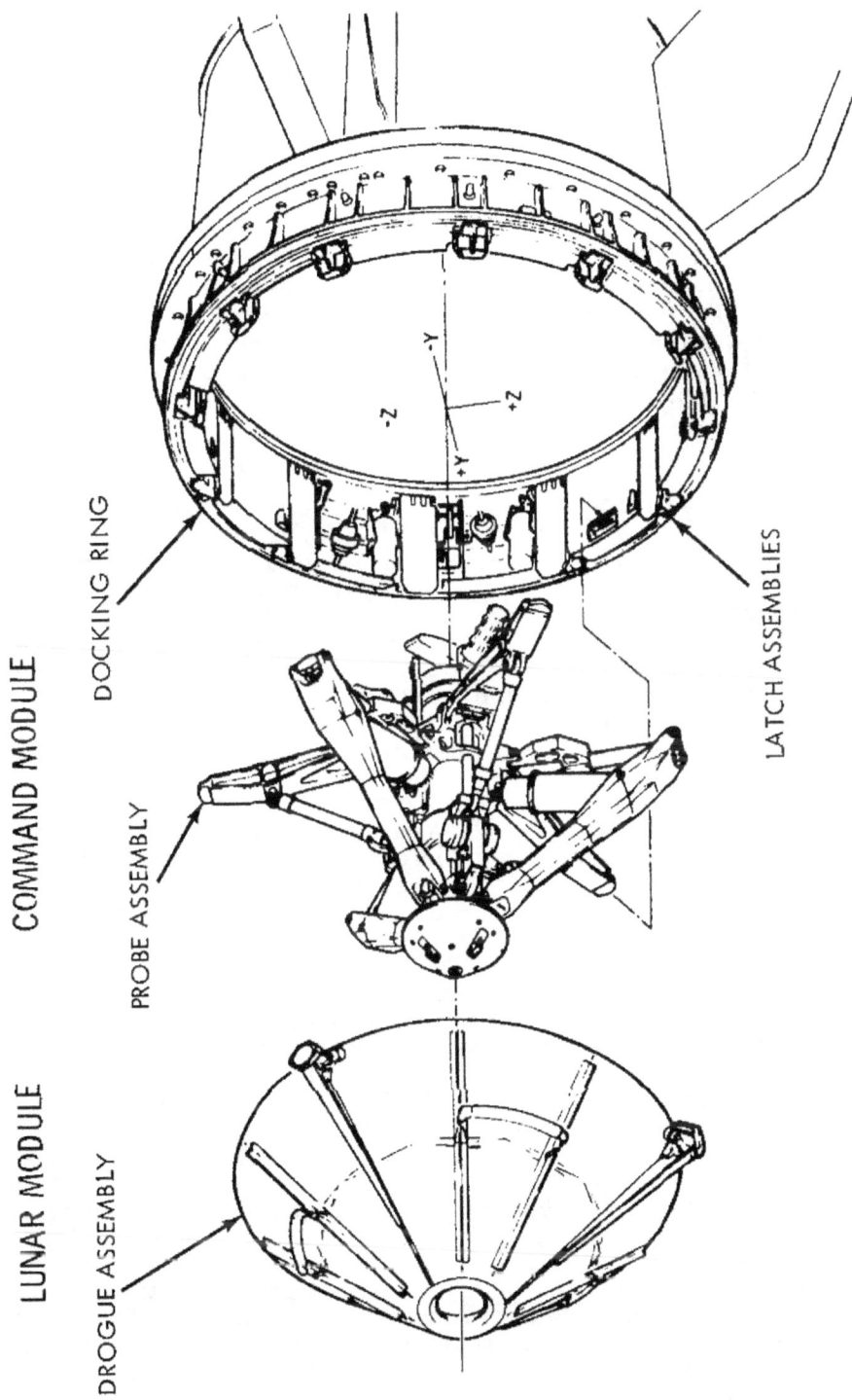

APOLLO DOCKING MECHANISMS

COMMAND MODULE

DOCKING RING

LATCH ASSEMBLIES

PROBE ASSEMBLY

LUNAR MODULE

DROGUE ASSEMBLY

-Y

-Z

+Z

+Y

Service Module (SM) Structure -- The service module is a
cylinder 12 feet 10 inches in diameter by 24 feet 7 inches high.
For the Apollo 10 mission, it will weigh, 51,371 pounds at launch.
Aluminum honeycomb panels one inch thick form the outer skin, and
milled aluminum radial beams separate the interior into six
sections containing service propulsion system and reaction control
fuel-oxidizer tankage, fuel cells, cryogenic oxygen and hydrogen,
and onboard consumables.

Spacecraft-LM Adapter (SLA) Structure -- The spacecraft LM
adapter is a truncated cone 28 feet long tapering from 260 inches
diameter at the base to 154 inches at the forward end at the
service module mating line. Aluminum honeycomb 1.75 inches thick
is the stressed-skin structure for the spacecraft adapter. The
SLA weighs 4,000 pounds.

CSM Systems

Guidance, Navigation and Control System (GNCS) -- Measures
and controls spacecraft position, attitude, and velocity, cal-
culates trajectory, controls spacecraft propulsion system thrust
vector, and displays abort data. The guidance system consists of
three subsystems: inertial, made up of an inertial measurement
unit and associated power and data components; computer which
processes information to or from other components; and optics,
including scanning telescope and sextant for celestial and/or
landmark spacecraft navigation. CSM 106 and subsequent modules
are equipped with a VHF ranging device as a backup to the LM
rendezvous radar.

Stabilization and Control System (SCS) -- Controls spacecraft rotation, translation, and thrust vector and provides
displays for crew-initiated maneuvers; backs up the guidance
system. It has three subsystems: attitude reference, attitude
control, and thrust vector control.

Service Propulsion System (SPS) -- Provides thrust for large
spacecraft velocity changes through a gimbal-mounted 20,500-
pound-thrust hypergolic engine using a nitrogen tetroxide oxidizer
and a 50-50 mixture of unsymmetrical dimethyl hydrazine and
hydrazine fuel. Tankage of this system is in the service module.
The system responds to automatic firing commands from the guidance and navigation system or to manual commands from the crew.
The engine provides a constant thrust rate. The stabilization and
control system gimbals the engine to fire through the spacecraft
center of gravity.

Reaction Control System (RCS) -- The command module and the
service module each has its own independent system. The SM RCS
has four identical RCS "quads mounted around the SM 90 degrees
apart. Each quad has four 100 pound-thrust engines, two fuel and
two oxidizer tanks and a helium pressurization sphere. The SM RCS
provides redundant spacecraft attitude control through cross-coupling
logic inputs from the stabilization and guidance systems.

Small velocity change maneuvers can also be made with the
SM RCS. The CM RCS consists of two independent six-engine subsystems of six 93 pound-thrust engines each. Both subsystems
are activated after CM separation from the SM: one is used for
spacecraft attitude control during entry. The other serves in
standby as a backup. Propellants for both CM and SM RCS are
monomethyl hydrazine fuel and nitrogen tetroxide oxidizer with
helium pressurization. These propellants are hypergolic, i.e.,
they burn spontaneously when combined without an igniter.

Electrical Power System (EPS) -- Consists of three, 31-
cell Bacon-type hydrogen-oxygen fuel cell power plants in the
service module which supply 28-volt DC power, three 28-volt DC
zinc-silver oxide main storage batteries in the command module
lower equipment bay, and three 115-200-volt 400 hertz three-
phase AC inverters powered by the main 28-volt DC bus. The
inverters are also located in the lower equipment bay. Cryogenic
hydrogen and oxygen react in the fuel cell stacks to provide
electrical power, potable water, and heat. The command module
main batteries can be switched to fire pyrotechnics in an
emergency. A battery charger restores selected batteries to
full strength as required with power from the fuel cells.

Environmental Control System (ECS) -- Controls spacecraft
atmosphere, pressure, and temperature and manages water. In
addition to regulating cabin and suit gas pressure, temperature
and humidity, the system removes carbon dioxide, odors and
particles, and ventilates the cabin after landing. It collects
and stores fuel cell potable water for crew use, supplies water
to the glycol evaporators for cooling, and dumps surplus water
overboard through the urine dump valve. Proper operating temperature of electronics and electrical equipment is maintained
by this system through the use of the cabin heat exchangers, the
space radiators, and the flycol evaporators.

Telecommunications System -- Provides voice, television telemetry, and command data and tracking and ranging between the spacecraft and Earth, between the command module and the lunar module
and between the spacecraft and the extravehicular astronaut. It
also provides intercommunications between astronauts. The telecommunications system consists of pulse code modulated telemetry
for relaying to Manned Space Flight Network stations data on
spacecraft systems and crew condition, VHF/AM voice, and unified
S-Band tracking transponder, air-to-ground voice communications,
onboard television, and a VHF recovery beacon. Network stations
can transmit to the spacecraft such items as updates to the
Apollo guidance computer and central timing equipment, and real-
time commands for certain onboard functions.

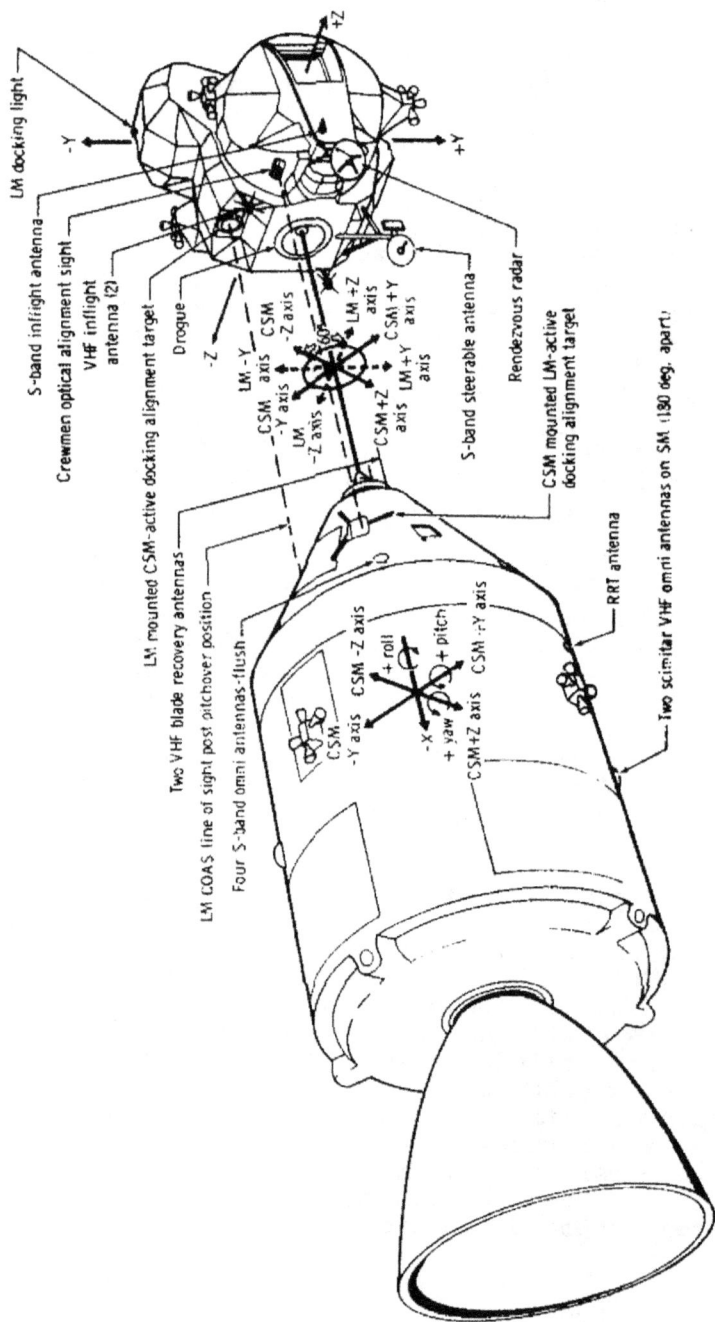

SPACECRAFT AXIS AND ANTENNA LOCATIONS

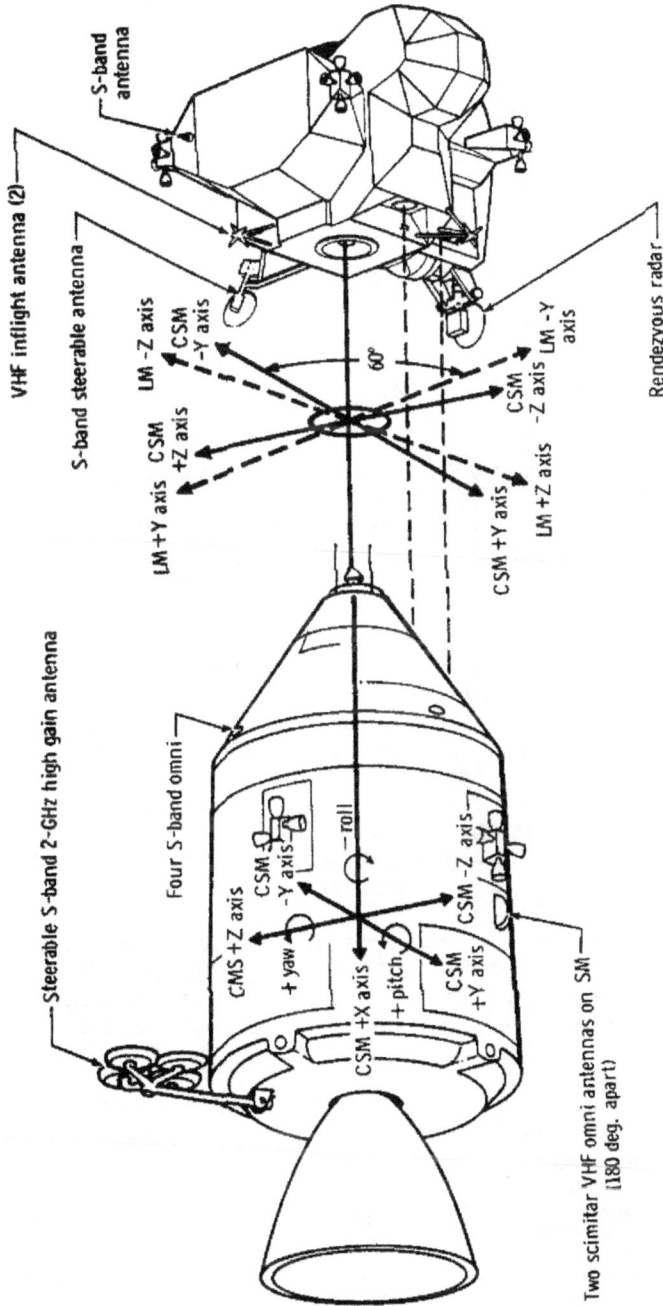

SPACECRAFT AXIS AND ANTENNA LOCATIONS

S-band antenna

VHF inflight antenna (2)

S-band steerable antenna

LM -Z axis
CSM -Y axis
CSM +Z axis
LM +Y axis

60°

CSM -Z axis LM -Y axis
CSM +Y axis LM +Z axis

Rendezvous radar

Steerable S-band 2-GHz high gain antenna

Four S-band omni

CMS +Z axis
+ yaw

CSM -Y axis
-roll

CSM -Z axis

CSM +X axis
+ pitch

CSM +Y axis

Two scimitar VHF omni antennas on SM
(180 deg. apart)

The high-gain steerable S-Band antenna consists of four, 31-inch-diameter parabolic dishes mounted on a folding boom at the aft end of the service module. Nested alongside the service propulsion system engine nozzle until deployment, the antenna swings out at right angles to the spacecraft longitudinal axis, with the boom pointing 52 degrees below the heads-up horizontal. Signals from the ground stations can be tracked either automatically or manually with the antenna's gimballing system. Normal S-Band voice and uplink/downlink communications will be handled by the omni and high-gain antennas.

Sequential System -- Interfaces with other spacecraft systems and subsystems to initiate time critical functions during launch, docking maneuvers, sub-orbital aborts, and entry portions of a mission. The system also controls routine spacecraft sequencing such as service module separation and deployment of the Earth landing system.

Emergency Detection System (EDS) -- Detects and displays to the crew launch vehicle emergency conditions, such as excessive pitch or roll rates or two engines out, and automatically or manually shuts down the booster and activates the launch escape system; functions until the spacecraft is in orbit.

Earth Landing System (ELS) -- Includes the drogue and main parachute system as well as post-landing recovery aids. In a normal entry descent, the command module forward heat shield is jettisoned at 24,000 feet, permitting mortar deployment of two reefed 16.5-foot diameter drogue parachutes for orienting and decelerating the spacecraft. After disreef and drogue release, three pilot mortar deployed chutes pull out the three main 83.3-foot diameter parachutes with two-stage reefing to provide gradual inflation in three steps. Two main parachutes out of three can provide a safe landing.

Recovery aids include the uprighting system, swimmer interphone connections, sea dye marker, flashing beacon, VHF recovery beacon, and VHF transceiver. The uprighting system consists of three compressor-inflated bags to upright the spacecraft if it should land in the water apex down (stable II position).

Caution and Warning System -- Monitors spacecraft systems for out-of-tolerance conditions and alerts crew by visual and audible alarms so that crewmen may trouble-shoot the problem.

Controls and Displays -- Provide readouts and control functions of all other spacecraft systems in the command and service modules. All controls are designed to be operated by crewmen in pressurized suits. Displays are grouped by system and located according to the frequency the crew refers to them.

LUNAR MODULE STRUCTURES, WEIGHT

The lunar module is a two-stage vehicle designed for space operations near and on the Moon. The LM is incapable of reentering the atmosphere. The lunar module stands 22 feet 11 inches high and is 31 feet wide (diagonally across landing gear).

Joined by four explosive bolts and umbilicals, the ascent and descent stages of the LM operate as a unit until staging, when the ascent stage functions as a single spacecraft for rendezvous and docking with the CSM.

Ascent Stage

Three main sections make up the ascent stage: the crew compartment, midsection, and aft equipment bay. Only the crew compartment and midsection are pressurized (4.8 psig; 337.4 gm/sq cm) as part of the LM cabin; all other sections of the LM are unpressurized. The cabin volume is 235 cubic feet (6.7 cubic meters). The ascent stage measures 12 feet 4 inches high by 14 feet 1 inch in diameter.

Structurally, the ascent stage has six substructural areas: crew compartment, midsection, aft equipment bay, thrust chamber assembly cluster supports, antenna supports and thermal and micrometeoroid shield.

The cylindrical crew compartment is a semimonocoque structure of machined longerons and fusion-welded aluminum sheet and is 92 inches (2.35 m) in diameter and 42 inches (1.07 m) deep. Two flight stations are equipped with control and display panels, armrests, body restraints, landing aids, two front windows, an overhead docking window, and an alignment optical telescope in the center between the two flight stations. The habitable volume is 160 cubic feet.

Two triangular front windows and the 32-inch (0.81 m) square inward-opening forward hatch are in the crew compartment front face.

External structural beams support the crew compartment and serve to support the lower interstage mounts at their lower ends. Ring-stiffened semimonocoque construction is employed in the midsection, with chem-milled aluminum skin over fusion-welded longerons and stiffeners. Fore-and-aft beams across the top of the midsection join with those running across the top of the cabin to take all ascent stage stress loads and, in effect, isolate the cabin from stresses.

-more-

DOCKING WINDOW

DOCKING DROGUE ASSEMBLY

VHF ANTENNA

DOCKING TARGET

S-BAND STEERABLE ANTENNA

RENDEZVOUS RADAR ANTENNA

-BAND IN-FLIGHT ANTENNA (2)

AFT EQUIPMENT BAY

RCS THRUST CHAMBER ASSEMBLY CLUSTER (4)

WINDOWS (2)

C-BAND ANTENNA (2)

FLASH HEAD

FORWARD HATCH

DOCKING LIGHT (4)

LANDING GEAR

LANDING PAD

FORWARD

+z

LADDER

EGRESS PLATFORM

DESCENT ENGINE SKIRT

LUNAR LANDING ANTENNA

LUNAR SURFACE SENSING PROBE (4)

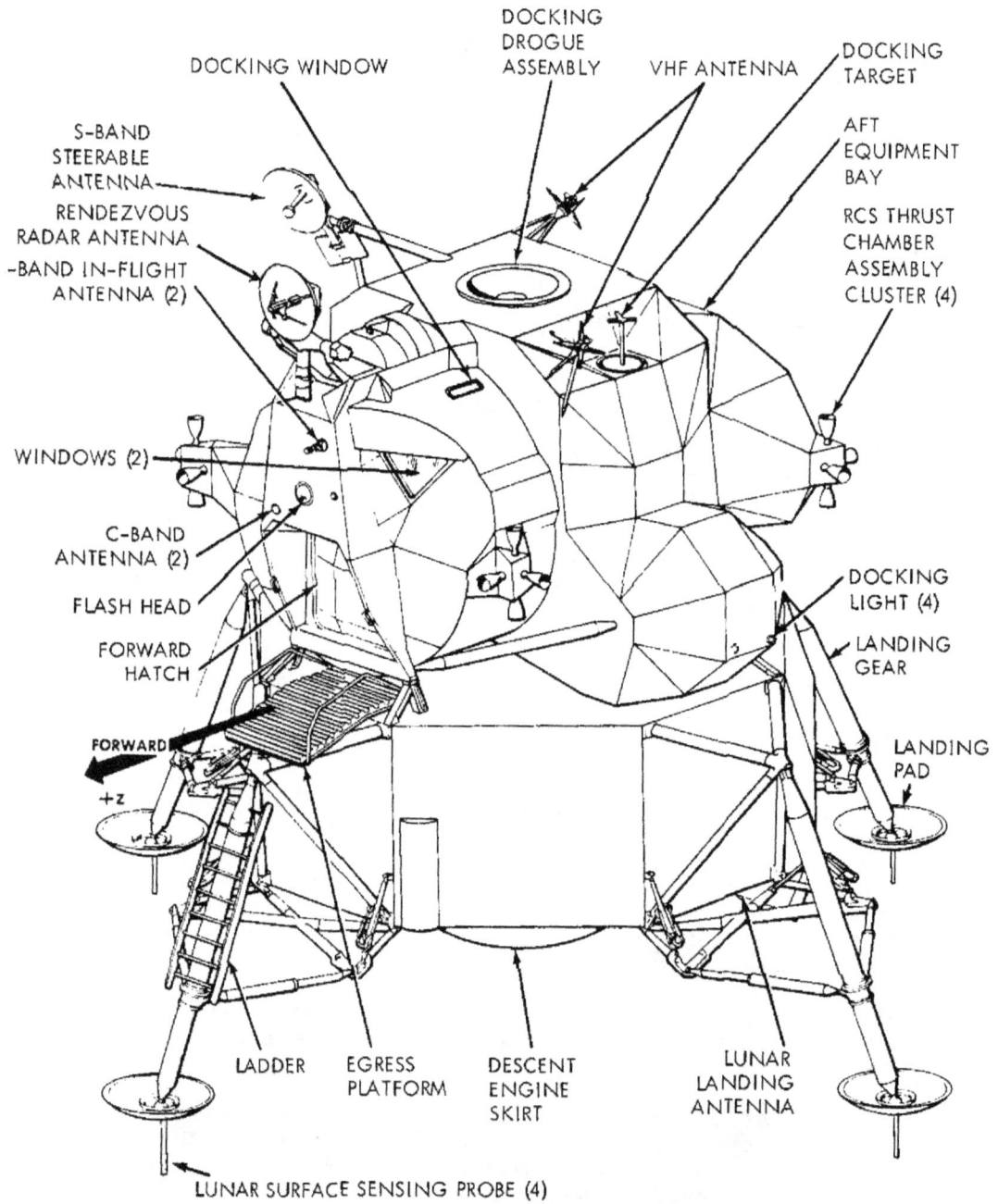

APOLLO LUNAR MODULE

-more-

VHF ANTENNA(2)

TRANSFER TUNNEL AND HATCH

DOCKING TARGET RECESS

GASEOUS OXYGEN TANK (2)

AFT EQUIPMENT BAY

REPLACEABLE ELECTRONIC ASSEMBLY

FUEL TANK (REACTION CONTROL)

LIQUID OXYGEN TANK

HELIUM TANK (2)

HELIUM TANK (REACTION CONTROL)

OXIDIZER TANK (REACTION CONTROL)

S-BAND STEERABLE ANTENNA

ALIGNMENT OPTICAL TELESCOPE

RENDEZVOUS RADAR ANTENNA

ASCENT ENGINE COVER

S-BAND INFLIGHT ANTENNA (2)

REACTION CONTROL ASSEMBLY (4 PLACES)

OXIDIZER TANK

INGRESS/EGRESS HATCH

CREW COMPARTMENT

WATER TANK(2)

FUEL TANK

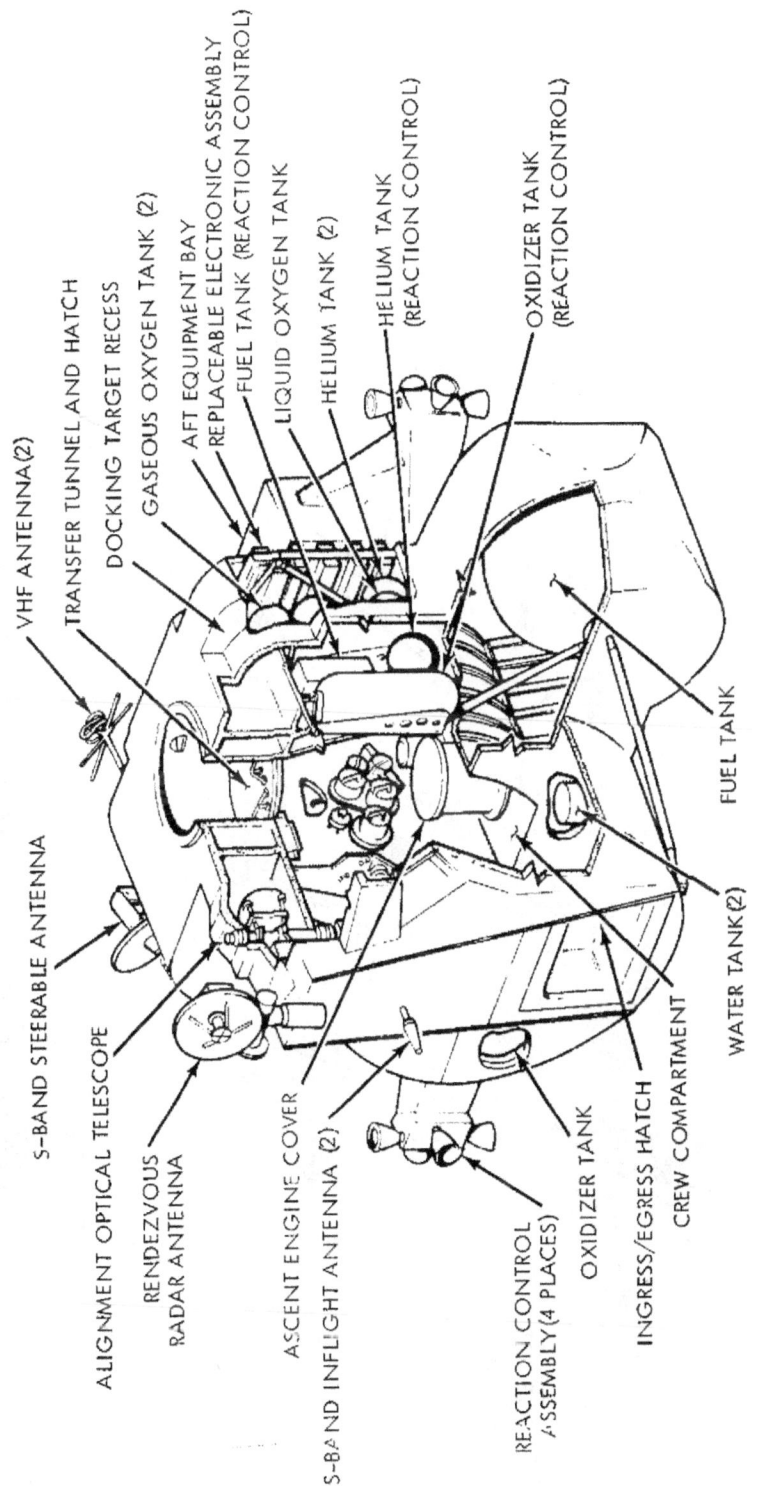

APOLLO LUNAR MODULE – ASCENT STAGE

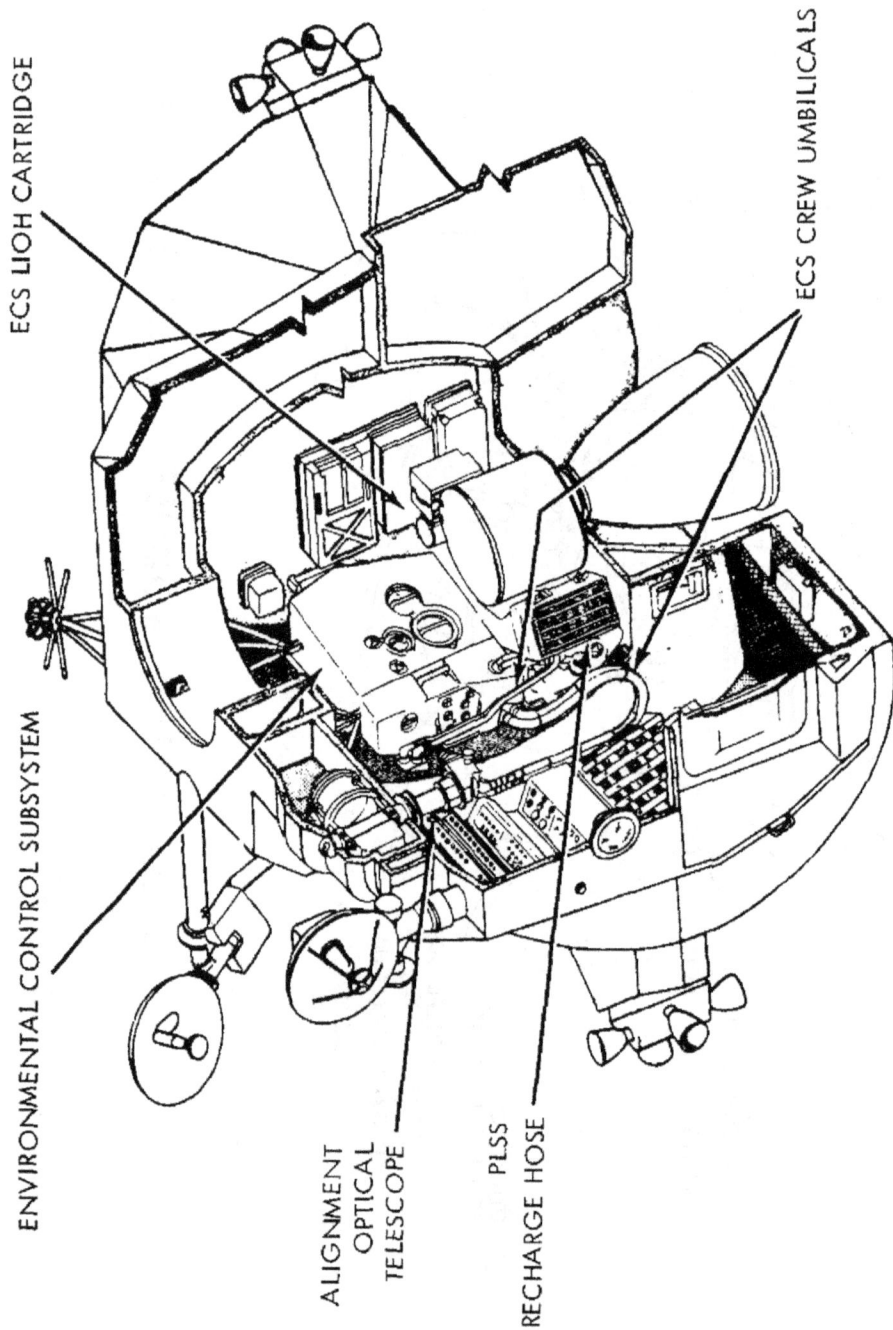

ECS LIOH CARTRIDGE

ECS CREW UMBILICALS

ENVIRONMENTAL CONTROL SUBSYSTEM

ALIGNMENT OPTICAL TELESCOPE

PLSS RECHARGE HOSE

LM CABIN INTERIOR, LEFT HALF

PLSS RECHARGE AND STOWAGE POSITION

PLSS O2 RECHARGE HOSE

DSEA

URINE MGT SYSTEM

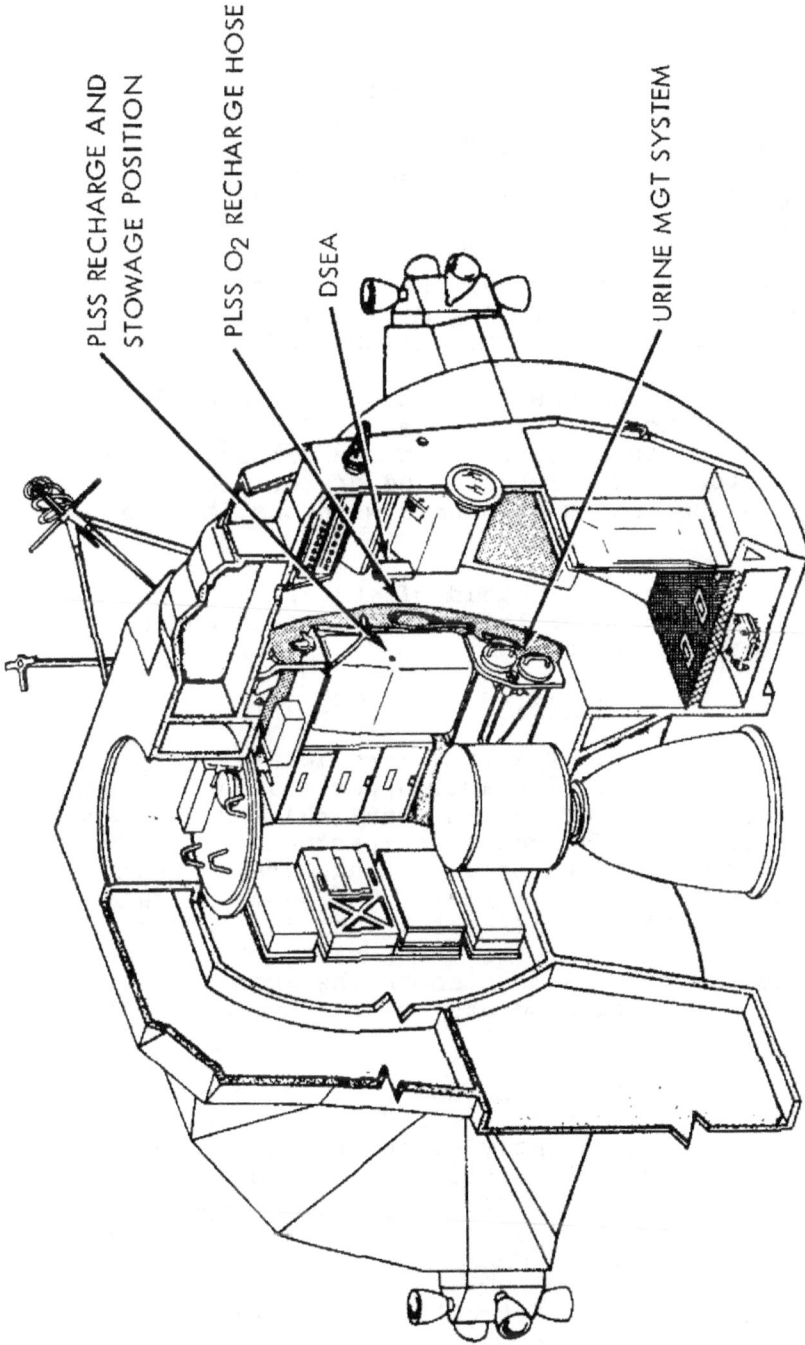

LM CABIN INTERIOR, RIGHT HALF

The ascent stage engine compartment is formed by two beams running across the lower midsection deck and mated to the fore and aft bulkheads. Systems located in the midsection include the LM guidance computer, the power and servo assembly, ascent engine propellant tanks, RCS propellant tanks, the environmental control system, and the waste management section.

A tunnel ring atop the ascent stage meshes with the command module latch assemblies. During docking, the ring and clamps are aligned by the LM drogue and the CSM probe.

The docking tunnel extends downward into the midsection 16 inches (40 cm). The tunnel is 32 inches (0.81 cm) in diameter and is used for crew transfer between the CSM and LM by crewmen. The upper hatch on the inboard end of the docking tunnel hinges downward and cannot be opened with the LM pressurized and undocked.

A thermal and micrometeoroid shield of multiple layers of mylar and a single thickness of thin aluminum skin encases the entire ascent stage structure.

Descent Stage

The descent stage consists of a cruciform load-carrying structure of two pairs of parallel beams, upper and lower decks, and enclosure bulkheads -- all of conventional skin-and-stringer aluminum alloy construction. The center compartment houses the descent engine, and descent propellant tanks are housed in the four square bays around the engine. The descent stage measures 10 feet 7 inches high by 14 feet 1 inch in diameter.

Four-legged truss outriggers mounted on the ends of each pair of beams serve as SLA attach points and as "knees" for the landing gear main struts.

Triangular bays between the main beams are enclosed into quadrants housing such components as the ECS water tank, helium tanks, descent engine control assembly of the guidance, navigation and control subsystem, ECS gaseous oxygen tank, and batteries for the electrical power system. Like the ascent stage, the descent stage is encased in the mylar and aluminum alloy thermal and micrometeoroid shield.

The LM external platform, or "porch", is mounted on the forward outrigger just below the forward hatch. A ladder extends down the forward landing gear strut from the porch for crew lunar surface operations.

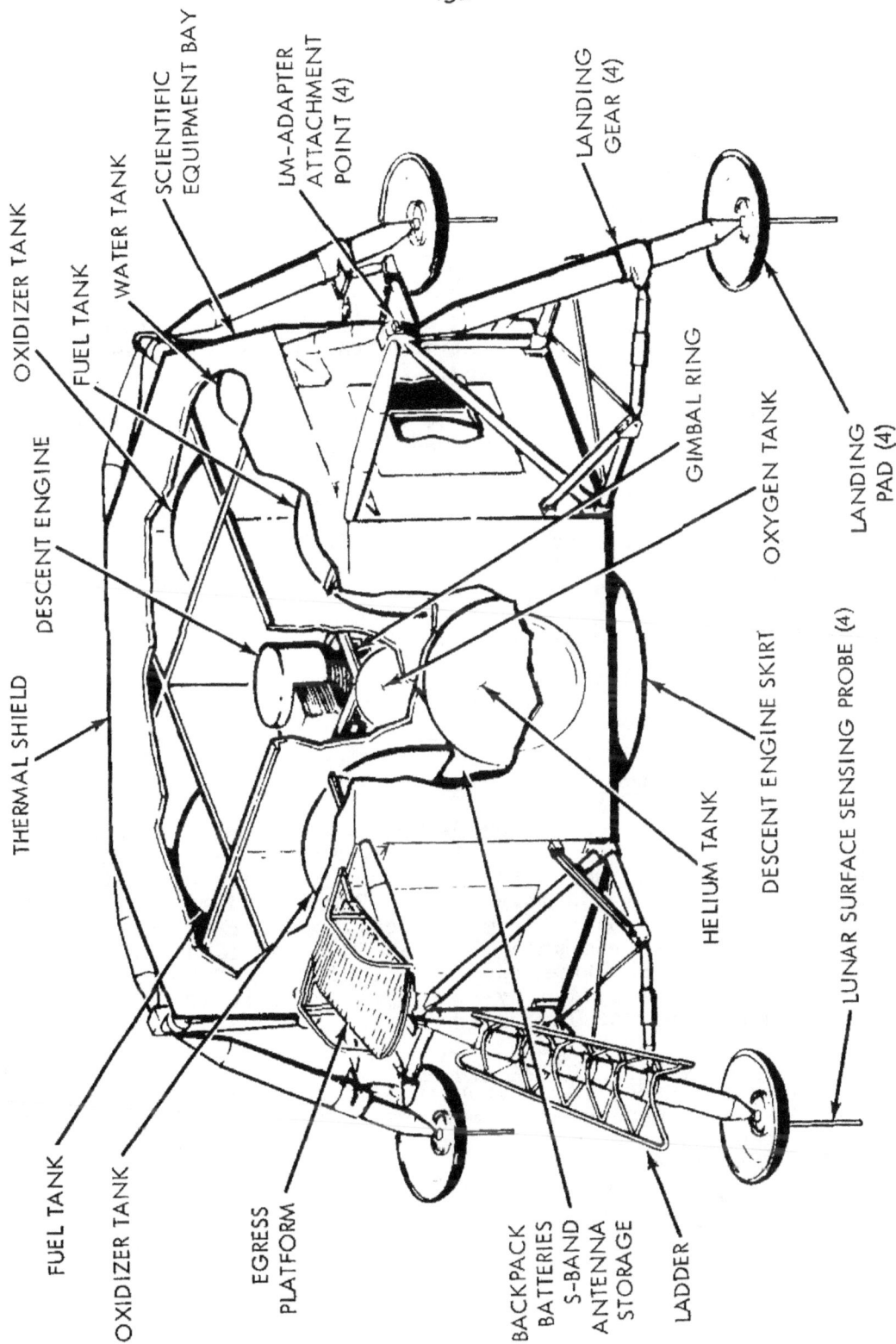

In a retracted position until after the crew mans the
LM, the landing gear struts are explosively extended and
provide lunar surface landing impact attenuation. The main
struts are filled with crushable aluminum honeycomb for
absorbing compression loads. Footpads 37 inches (0.95 m) in
diameter at the end of each landing gear provide vehicle
"floatation" on the lunar surface.

Each pad is fitted with a lunar-surface sensing probe
which signals the crew to shut down the descent engine upon
contact with the lunar surface.

LM-4 flown on the Apollo 10 mission will have a launch
weight of 30,849 pounds. The weight breakdown is as follows:

Ascent stage, dry	4,781 lbs.
Descent stage, dry	4,703 lbs.
RCS propellants	612 lbs.
DPS propellants	18,134 lbs.
APS propellants	2,619 lbs.
	30,849 lbs.

Lunar Module Systems

Electrical Power System -- The LM DC electrical system
consists of six silver zinc primary batteries -- four in the
descent stage and two in the ascent stage, each with its own
electrical control assembly (ECA). Power feeders from all
primary batteries pass through circuit breakers to energize
the LM DC buses, from which 28-volt DC power id distributed
through circuit breakers to all LM systems. AC power
(117v 400Hz) is supplied by two inverters, either of which can
supply spacecraft AC load needs to the AC buses.

Environmental Control System -- Consists of the atmosphere
revitalization section, oxygen supply and cabin pressure control
section, water management, heat transport section, and outlets
for oxygen and water servicing of the Portable Life Support
System (PLSS).

Components of the atmosphere revitalization section are the suit circuit assembly which cools and ventilates the pressure garments, reduces carbon dioxide levels, removes odors, noxious gases and excessive moisture; the cabin recirculation assembly which ventilates and controls cabin atmosphere temperatures; and the steam flex duct which vents to space steam from the suit circuit water evaporator.

The oxygen supply and cabin pressure section supplies gaseous oxygen to the atmosphere revitalization section for maintaining suit and cabin pressure. The descent stage oxygen supply provides descent flight phase and lunar stay oxygen needs, and the ascent stage oxygen supply provides oxygen needs for the ascent and rendezvous flight phase.

Water for drinking, cooling, fire fighting, food preparation, and refilling the PLSS cooling water servicing tank is supplied by the water management section. The water is contained in three nitrogen-pressurized bladder-type tanks, one of 367-pound capacity in the descent stage and two of 47.5-pound capacity in the ascent stage.

The heat transport section has primary and secondary water-glycol solution coolant loops. The primary coolant loop circulates water-glycol for temperature control of cabin and suit circuit oxygen and for thermal control of batteries and electronic components mounted on cold plates and rails. If the primary loop becomes inoperative, the secondary loop circulates coolant through the rails and cold plates only. Suit circuit cooling during secondary coolant loop operation is provided by the suit loop water boiler. Waste heat from both loops is vented overboard by water evaporation or sublimators.

Communication System -- Two S-band transmitter-receivers, two VHF transmitter-receivers, a signal processing assembly, and associated spacecraft antenna make up the LM communications system. The system transmits and receives voice, tracking and ranging data, and transmits telemetry data on 281 measurements and TV signals to the ground. Voice communications between the LM and ground stations is by S-band, and between the LM and CSM voice is on VHF.

Real-time commands to the lunar module are received and encoded by the digital uplink assembly--a black box tied in to the S-band receiver. The digital uplink assembly will be used on Apollo 10 to arm and fire the ascent propulsion system for the unmanned APS depletion burn following final docking and LM jettison. LM-4 will be the last spacecraft to be fitted with equipment for accepting real-time commands from the ground.

The data storage electronics assembly (DSEA) is a four-channel voice recorder with timing signals with a 10-hour recording capacity which will be brought back into the CSM for return to Earth. DSEA recordings cannot be "dumped" to ground stations.

LM antennas are one 26-inch diameter parabolic S-band steerable antenna, two S-band inflight antennas and two VHF inflight antennas.

Guidance, Navigation and Control System -- Comprised of six sections: primary guidance and navigation section (PGNS), abort guidance section (AGS), radar section, control electronics section (CES), and orbital rate drive electronics for Apollo and LM (ORDEAL).

* The PGNS is an inertial system aided by the alignment optical telescope, an inertial measurement unit. and the rendezvous and landing radars. The system provides inertial reference data for computations, produces inertial alignment reference by feeding optical sighting data into the LM guidance computer, displays position and velocity data, computes LM-CSM rendezvous data from radar inputs, controls attitude and thrust to maintain desired LM trajectory, and controls descent engine throttling and gimbaling.

* The AGS is an independent backup system for the PGNS, having its own inertial sensor and computer.

* The radar section is made up of the rendezvous radar which provides CSM range and range rate, and line-of-sight angles for maneuver computation to the LM guidance computer; the landing radar which provide altitude and velocity data to the LM guidance computer during lunar landing. The rendezvous radar has an operating range from 80 feet to 400 nautical miles. The range transfer tone assembly, utilizing VHF electronics, is a passive responder to the CSM VHF ranging device and is a backup to the rendezvous radar.

* The CES controls LM attitude and translation about all axes. It also controls by PGNS command the automatic operation of the ascent and descent engines, and the reaction control thrusters. Manual attitude controller and thrust-translation controller commands are also handled by the CES.

* ORDEAL, displays on the flight director attitude indicator, is the computed local vertical in the pitch axis during circular, Earth or lunar orbits.

Reaction Control System -- The LM has four RCS engine clusters of four 100-pound (45.4 kg) thrust engines each which use helium-pressurized hypergolic propellants. The oxidizer is nitrogen tetroxide, fuel is Aerozine 50 (50/50 blend of hydrazine and unsymmetrical dimethyl hydrazine). Propellant plumbing, valves and pressurizing components are in two parallel, independent systems, each feeding half the engines in each cluster. Either system is capable of maintaining attitude alone, but if one supply system fails, a propellant crossfeed allows one system to supply all 16 engines. Additionally, interconnect valves permit the RCS system to draw from ascent engine propellant tanks.

The engine clusters are mounted on outriggers 90 degrees apart on the ascent stage.

The RCS provides small stabilizing impulses during ascent and descent burns, controls LM attitude during maneuvers, and produces thrust for separation, and ascent/descent engine tank ullage. The system may be operated in either the pulse or steady-state modes.

Descent Propulsion System -- Maximum rated thrust of the descent engine is 9,870 pounds (4,380.9 kg) and is throttleable between 1,050 pounds (476.7 kg) and 6,300 pounds (2,860.2 kg). The engine can be gimbaled six degrees in any direction for offset center of gravity trimming. Propellants are helium-pressurized Aerozine 50 and nitrogen tetroxide.

Ascent Propulsion System -- The 3,500-pound (1,589 kg) thrust ascent engine is not gimbaled and performs at full thrust. The engine remains dormant until after the ascent stage separates from the descent stage. Propellants are the same as are burned by the RCS engines and the descent engine.

Caution and Warning, Controls and Displays -- These two systems have the same function aboard the lunar module as they do aboard the command module. (See CSM systems section.)

Tracking and Docking Lights -- A flashing tracking light
(once per second, 20 milliseconds duration) on the front face
of the lunar module is an aid for contingency CSM-active
rendezvous LM rescue. Visibility ranges from 400 nautical
miles through the CSM sextant to 130 miles with the naked eye.
Five docking lights analagous to aircraft running lights are
mounted on the LM for CSM-active rendezvous: two forward
yellow lights, aft white light, port red light and starboard
green light. All docking lights have about a 1,000-foot
visibility.

SATURN V LAUNCH VEHICLE DESCRIPTION AND OPERATION

The Saturn V, 363 feet tall with the Apollo spacecraft in place, generates enough thrust to place a 125-ton payload into a 105-nm circular orbit of the Earth. It can boost about 50 tons to lunar orbit. The thrust of the three propulsive stages range from almost 7.6 million pounds for the booster to 230,000 pounds for the third stage at operating altitude. Including the instrument unit, the launch vehicle without the spacecraft is 281 feet tall.

First Stage

The first stage (S-IC) was developed jointly by the National Aeronautics and Space Administration's Marshall Space Flight Center, Huntsville, Ala. and the Boeing Co.

The Marshall Center assembled four S-IC stages: a structural test model, a static test version, and the first two flight stages. Subsequent flight stages are assembled by Boeing at the Michoud Assembly Facility, New Orleans. The S-IC stage destined for the Apollo 10 mission was the second flight booster static tested at the NASA-Mississippi Test Facility. The first S-IC test at MTF was on May 11, 1967, and the test of the second S-IC there -- the booster for Apollo 10 -- was completed Aug. 9, 1967. Earlier flight stages were static fired at the Marshall Center.

The S-IC stage boosts the space vehicle to an altitude of 35.8 nm at 50 nm downrange and increases the vehicle's velocity to 5,343 knots in 2 minutes 40 seconds of powered flight. It then separates and falls into the Atlantic Ocean about 351 nm downrange (30 degrees North latitude and 74 degree West longitude) about nine minutes after liftoff.

Normal propellant flow rate to the five F-1 engines is 29,522 pounds per second. Four of the engines are mounted on a ring, each 90 degrees from its neighbor. These four are gimballed to control the rocket's direction of flight. The fifth engine is mounted rigidly in the center.

Second Stage

The second stage (S-II), like the third stage, uses high performance J-2 engines that burn liquid oxygen and liquid hydrogen. The stage's purpose is to provide stage boost nearly to Earth orbit.

SATURN V LAUNCH VEHICLE

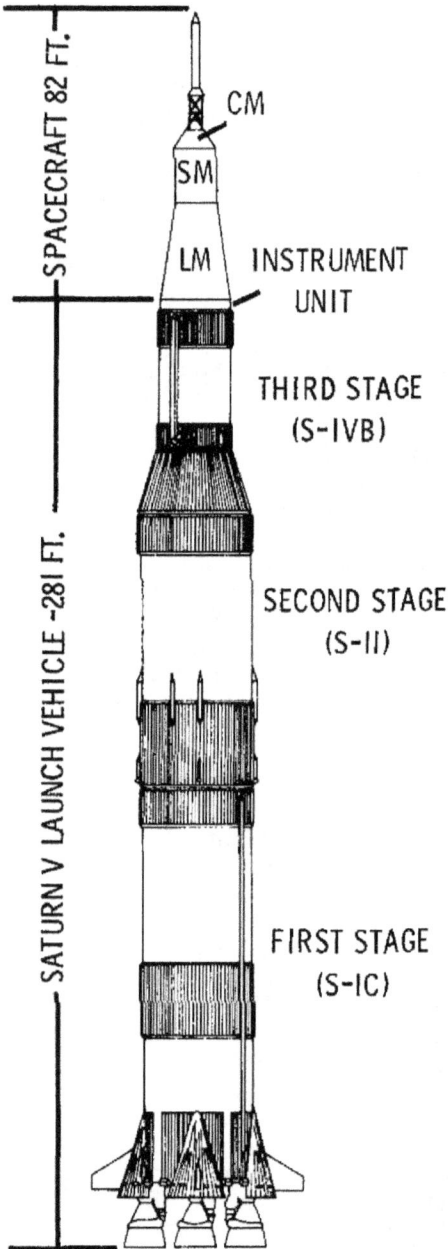

SPACECRAFT 82 FT.

CM

SM

LM — INSTRUMENT UNIT

THIRD STAGE (S-IVB)

SECOND STAGE (S-II)

SATURN V LAUNCH VEHICLE -281 FT.

FIRST STAGE (S-IC)

FIRST STAGE (S-IC)	
DIAMETER	33 FEET
HEIGHT	138 FEET
WEIGHT	5,031,023 LBS. FUELED 294,200 LBS .DRY
ENGINES	FIVE F-1
PROPELLANTS	LIQUID OXYGEN (3,258,280 LBS.) RP-I (KEROSENE) - (1,417,334 LBS.)
THRUST	7,680,982 LBS.

SECOND STAGE (S-II)	
DIAMETER	33 FEET
HEIGHT	81.5 FEET
WEIGHT	1,074,590 LBS. FUELED 84,367 LBS. DRY
ENGINES	FIVE J-2
PROPELLANTS	LIQUID OXYGEN (829,114 LBS.) LIQUID HYDROGEN (158,231 LBS.)
THRUST	1,163,854 LBS.
INTERSTAGE	8,890 LBS.

THIRD STAGE (S-IVB)	
DIAMETER	21.7 FEET
HEIGHT	58.3 FEET.
WEIGHT	261,836 LBS. FUELED 25,750 LBS. DRY
ENGINES	ONE J-2
PROPELLANTS	LIQUID OXYGEN (190,785 LBS.) LIQUID HYDROGEN (43,452 LBS.)
THRUST	203,615 LBS.
INTERSTAGE	8,081 LBS.

INSTRUMENT UNIT	
DIAMETER	21.7 FEET
HEIGHT	3 FEET
WEIGHT	4,254 LBS.

NOTE: WEIGHTS AND MEASURES GIVEN ABOVE ARE FOR THE NOMINAL VEHICLE CONFIGURATION FOR APOLLO 10. THE FIGURES MAY VARY SLIGHTLY DUE TO CHANGES BEFORE LAUNCH TO MEET CHANGING CONDITIONS.

-more-

At outboard engine cutoff, the S-II separates and, following a ballistic trajectory, plunges into the Atlantic Ocean about 2,400 nm downrange from Kennedy Space Center (31 degrees North latitude and 34 degrees West longitude) about 20 minutes after liftoff.

Five J-2 engines power the S-II. The outer four engines are equally spaced on a 17.5-foot diameter circle. These four engines may be gimbaled through a plus or minus seven-degree square pattern for thrust vector control. As on the first stage, the center engine (number 5) is mounted on the stage centerline and is fixed in position.

The S-II carries the rocket to an altitude of about 101.6 nm and a distance of some 888 nm downrange. Before burnout, the vehicle will be moving at a speed of 13,427 knots. The outer J-2 engines will burn 6 minutes 32 seconds during this powered phase, but the center engine will be cut off at 4 minutes 59 seconds of burn time.

The Space Division of North American Rockwell Corp. builds the S-II at Seal Beach, Calif. The cylindrical vehicle is made up of the forward skirt to which the third stage attaches, the liquid hydrogen tank, the liquid oxygen tank (separated from the hydrogen tank by a common bulkhead), the thrust structure on which the engines are mounted and an interstage section to which the first stage attaches. The common bulkhead between the two tanks is heavily insulated.

The S-II for Apollo 10 was static tested by North American Rockwell at the NASA-Mississippi Test Facility on Aug. 9, 1968. This stage was shipped to the test site via the Panama Canal for the test firing.

Third Stage

The third stage (S-IVB) was developed by the McDonnell Douglas Astronautics Co. at Huntington Beach, Calif. At Sacramento, Calif., the stage passed a static firing test on Oct. 9, 1967 as part of the preparation for the Apollo 10 mission. The stage was flown directly to the NASA-Kennedy Space Center.

Measuring 58 feet 4 inches long and 21 feet 8 inches in diameter, the S-IVB weighs 25,750 pounds dry. At first ignition, it weighs 261,836 pounds. The interstage section weighs an additional 8,081 pounds. The stage's J-2 engine burns liquid oxygen and liquid hydrogen.

The stage provides propulsion twice during the Apollo
10 mission. The first burn occurs immediately after separation from the S-II. It will last long enough (156 seconds)
to insert the vehicle and spacrcraft into a circular Earth
parking orbit at about 52 degrees West longitude and 32 degrees
North latitude.

The second burn, which begins at 2 hours 33 minutes 25
seconds after liftoff (for first opportunity translunar injection) or 4 hours 2 minutes 5 seconds (for second TLI opportunity), will place the stage, instrument unit, and spacecraft
into translunar trajectory. The burn will continue until
proper TLI end conditions are met.

The fuel tanks contain 43,452 pounds of liquid hydrogen
and 190,785 pounds of liquid oxygen at first ignition, totalling
234,237 pounds of propellants. Insulation between the two
tanks is necessary because the liquid oxygen, at about 293
degrees below zero F, is warm enough, relatively, to heat the
liquid hydrogen, at 423 degrees below zero F, rapidly and cause
it to turn into gas.

Instrument Unit

The instrument unit (IU) is a cylinder three feet high
and 21 feet 8 inches in diameter. It weighs 4,254 pounds and
contains the guidance, navigation, and control equipment which
will steer the vehicle through its Earth orbits and into the
final translunar injection maneuver.

The IU also contains telemetry, communications, tracking,
and crew safety systems, along with its own supporting electrical
power and environmental control systems.

Components making up the "brain" of the Saturn V are
mounted on cooling panels fastened to the inside surface of
the instrument unit skin. The "cold plates" are part of a
system that removes heat by circulating cooled fluid through
a heat exchanger that evaporates water from a separate supply
into the vacuum of space.

The six major systems of the instrument unit are
structural, thermal control, guidance and control, measuring
and telemetry, radio frequency, and electrical.

The instrument unit provides navigation, guidance,
and control of the vehicle; measurement of vehicle performance
and environment; data transmission with ground stations; radio
tracking of the vehicle; checkout and monitoring of vehicle
functions; initiation of stage functional sequencing; detection
fo emergency situations; generation and network distribution of
electric power system operation; and preflight checkout and
launch and flight operations.

A path-adaptive guidance scheme is used in the Saturn V instrument unit. A programmed trajectory is used in the initial launch phase with guidance beginning only after the vehicle has left the atmosphere. This is to prevent movements that might cause the vehicle to break apart while attempting to compensate for winds, jet streams, and gusts encountered in the atmosphere.

If such air currents displace the vehicle from the optimum trajectory in climb, the vehicle derives a new trajectory. Calculations are made about once each second throughout the flight. The launch vehicle digital computer and data adapter perform the navigation and guidance computations.

The ST-124M inertial platform -- the heart of the navigation, guidance and control system -- provides space-fixed reference coordinates and measures acceleration along the three mutually perpendicular axes of the coordinate system.

International Business Machines Corp., is prime contractor for the instrument unit and is the supplier of the guidance signal processor and guidance computer. Major suppliers of instrument unit components are: Electronic Communications, Inc., control computer; Bendix Corp., ST-124M inertial platform; and IBM Federal Systems Division, launch vehicle digital computer and launch vehicle data adapter.

Propulsion

The 41 rocket engines of the Saturn V have thrust ratings ranging from 72 pounds to more than 1.5 million pounds. Some engines burn liquid propellants, others use solids.

The five F-1 engines in the first stage burn RP-1 (kerosene) and liquid oxygen. Engines in the first stage develop approximately 1,536,197 pounds of thrust each at liftoff, building up to 1,822,987 pounds before cutoff. The cluster of five engines gives the first stage a thrust range from 7,680,982 million pounds at liftoff to 9,114,934 pounds just before center engine cutoff.

The F-1 engine weighs almost 10 tons, is more than 18 feet high and has a nozzle-exit diameter of nearly 14 feet. The F-1 undergoes static testing for an average 650 seconds in qualifying for the 160-second run during the Saturn V first stage booster phase. The engine consumes almost three tons of propellants per second.

The first stage of the Saturn V for this mission has eight other rocket motors. These are the solid-fuel retro-rockets which will slow and separate the stage from the second stage. Each rocket produces a thrust of 87,900 pounds for 0.6 second.

The main propulsion for the second stage is a cluster of five J-2 engines burning liquid hydrogen and liquid oxygen. Each engine develops a mean thrust of more than 205,000 pounds at 5.0:1 mixture ratio (variable from 184,000 to 230,000 in phases of flight), giving the stage a total mean thrust of more than a million pounds.

Designed to operate in the hard vacuum of space, the 3,500-pound J-2 is more efficient than the F-1 because it burns the high-energy fuel hydrogen. F-1 and J-2 engines are produced by the Rocketdyne Division of North American Rockwell Corp.

The second stage has four 21,000-pound-thrust solid fuel rocket engines. These are the ullage rockets mounted on the S-IC/S-II interstage section. These rockets fire to settle liquid propellant in the bottom of the main tanks and help attain a "clean" separation from the first stage, they remain with the interstage when it drops away at second plane separation. Four retrorockets are located in the S-IVB aft interstage (which never separates from the S-II) to separate the S-II from the S-IVB prior to S-IVB ignition.

Eleven rocket engines perform various functions on the third stage. A single J-2 provides the main propulsive force; there are two jettisonable main ullage rockets and eight smaller engines in the two auxiliary propulsion system modules.

Launch Vehicle Instrumentation and Communication

A total of 2,342 measurements will be taken in flight on the Saturn V launch vehicle: 672 on the first stage, 986 on the second stage, 386 on the third stage, and 298 on the instrument unit.

The Saturn V has 16 telemetry systems: six on the first stage, six on the second stage, one on the third stage and three on the instrument unit. A C-band system and command system are also on the instrument unit. Each powered stage has a range safety system as on previous flights.

S-IVB Restart

The third stage of the Saturn V rocket for the Apollo 10 mission will burn twice in space. The second burn places the spacecraft on the translunar trajectory. The first opportunity for this burn is at 2 hours 33 minutes and 25 seconds after launch.. The second opportunity for TLI begins at 4 hours 2 minutes and 5 seconds after liftoff.

The primary pressurization system of the propellant tanks for the S-IVB restart uses a helium heater. In this sytem, nine helium storage spheres in the liquid hydrogen tank contain gaseous helium charged to about 3,000 psi. This helium is passed through the heater which heats and expands the gas before it enters the propellant tanks. The heater operates on hydrogen and oxygen gas from the main propellant tanks.

The backup system consists of five ambient helium spheres mounted on the stage thrust structure. This system, controlled by the fuel repressurization control module, can repressurize the tanks in case the primary system fails. The restart will use the primary system. If that system fails, the backup system will be used.

The third stage for Apollo 10 will not be ignited for a third burn as on Apollo 9. Following spacecraft separation in translunar trajectory, the stage will undergo the normal J-2 engine chilldown sequence, stopping just short of reignition. On Apollo 10 there is no requirement for a third burn, and there will not be sufficient propellants aboard, most of the fuels having been expended during the translunar injection maneuver.

Differences in Apollo 9 and Apollo 10 Launch Vehicles

Two modifications resulting from problems encountered during the second Saturn V flight were incorporated and proven successful on the third and fourth Saturn V missions. The new helium prevalve cavity pressurization system will again be flown on the first (S-IC) stage of Apollo 10. New augmented spark igniter lines which flew on the engines of the two upper stages of Apollo 8 and 9 will again be used on Apollo 10.

The major first stage (S-IC) differences between Apollo 9 and 10 are:

1. Dry weight was reduced from 295,600 to 294,200 pounds.

2. Weight at ground ignition increased from 5,026,200 to 5,031,023 pounds.

3. Instrumentation measurements were increased from 666 to 672.

S-II stage changes are:

1. Nominal vacuum thrust for J-2 engines increase will change maximum stage thrust from 1,150,000 to 1,168,694 pounds.

2. The approximate empty weight of the S-II has been reduced from 84,600 to 84,367 pounds. The S-IC/S-II interstage weight was reduced from 11,664 to 8,890 pounds.

3. Approximate stage gross liftoff weight was increased from 1,069,114 to 1,074,590 pounds.

4. Instrumentation measurements increased from 975 to 986.

Major differences on the S-IVB stage of Apollo 9 and 10 are:

1. S-IVB dry stage weight increased from 25,300 to 25,750 pounds. This does not include the 8,084-pound interstage section.

2. S-IVB gross stage weight at liftoff increased from 259,337 to 261,836 pounds.

3. Instrumentation measurements were increased from 296 to 386.

APOLLO 10 CREW

Life Support Equipment - Space Suits

Apollo 10 crewmen will wear two versions of the Apollo space suit: an intravehicular pressure garment assembly worn by the command module pilot and the extravehicular pressure garment assembly worn by the commander and the lunar module pilot. Both versions are basically identical except that the extravehicular version has an integral thermal/meteoroid garment over the basic suit.

From the skin out, the basic pressure garment consists of a nomex comfort layer, a neoprene-coated nylon pressure bladder and a nylon restraint layer. The outer layers of the intravehicular suit are, from the inside out, nomex and two layers of Teflon-coated Beta cloth. The extravehicular integral thermal/meteoroid cover consists of a liner of two layers of neoprene-coated nylon, seven layers of Beta/Kapton spacer laminate, and an outer layer of Teflon-coated Beta fabric.

The extravehicular suit, together with a liquid cooling garment, portable life support system (PLSS), oxygen purge system, extravehicular visor assembly and other components make up the extravehicular mobility unit (EMU). The EMU provides an extravehicular crewman with life support for a four-hour mission outside the lunar module without replenishing expendables. EMU total weight is 183 pounds. The intravehicular suit weighs 35.6 pounds.

Liquid cooling garment--A knitted nylon-spandex garment with a network of plastic tubing through which cooling water from the PLSS is circulated. It is worn next to the skin and replaces the constant wear-garment during EVA only.

Portable life support system--A backpack supplying oxygen at 3.9 psi and cooling water to the liquid cooling garment. Return oxygen is cleansed of solid and gas contaminants by a lithium hydroxide canister. The PLSS includes communications and telemetry equipment, displays and controls, and a main power supply. The PLSS is covered by a thermal insulation jacket. (One stowed in LM).

Oxygen purge system--Mounted atop the PLSS, the oxygen purge system provides a contingency 30-minute supply of gaseous oxygen in two two-pound bottles pressurized to 5,880 psia. The system may also be worn separately on the front of the pressure garment assembly torso. It serves as a mount for the VHF antenna for the PLSS. (Two stowed in LM).

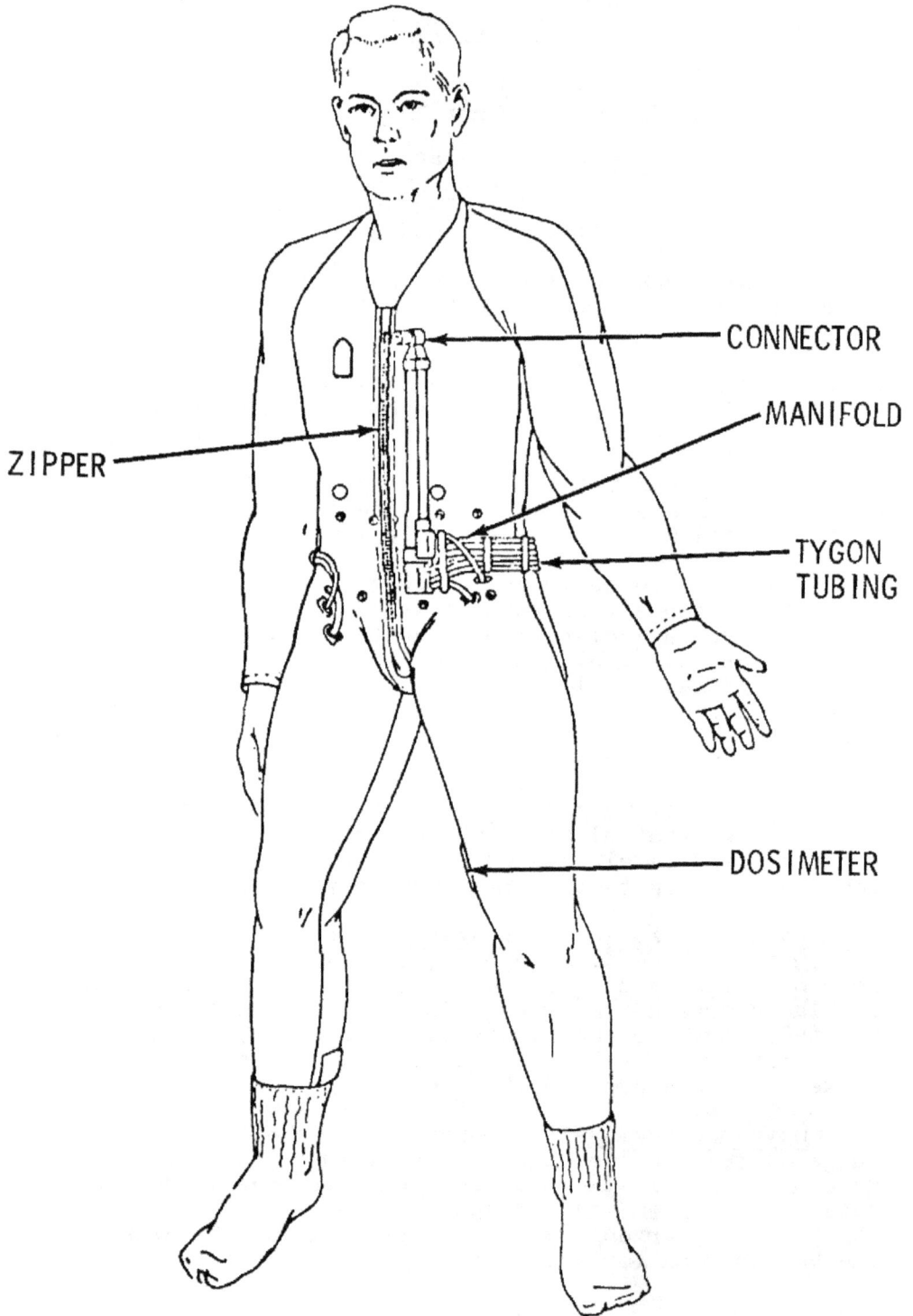

CONNECTOR

MANIFOLD

ZIPPER

TYGON
TUBING

DOSIMETER

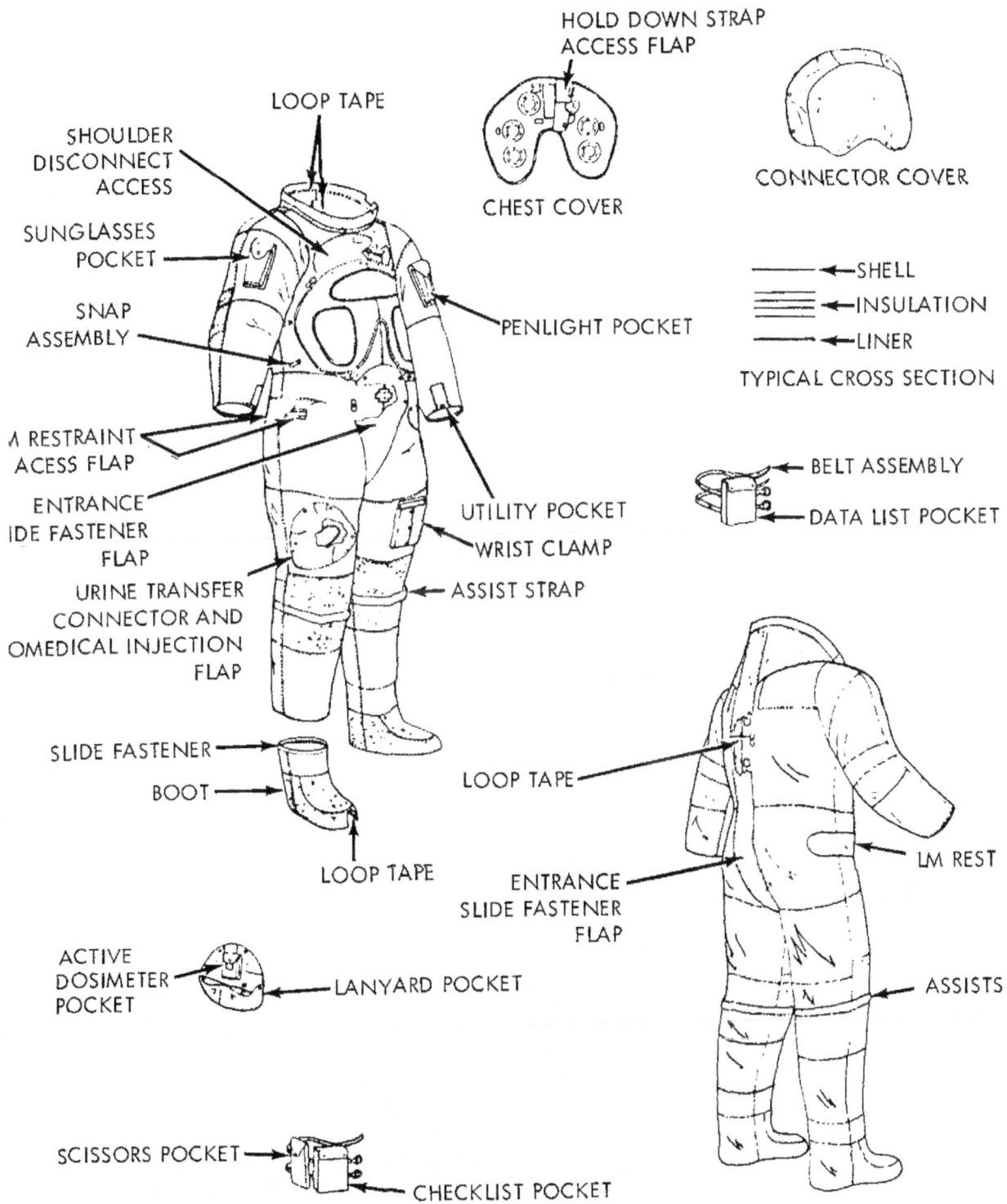

LOOP TAPE

HOLD DOWN STRAP
ACCESS FLAP

SHOULDER
DISCONNECT
ACCESS

CONNECTOR COVER

SUNGLASSES
POCKET

CHEST COVER

SNAP
ASSEMBLY

PENLIGHT POCKET

←SHELL

←INSULATION

←LINER

TYPICAL CROSS SECTION

A RESTRAINT
ACESS FLAP

ENTRANCE
IDE FASTENER
FLAP

UTILITY POCKET

WRIST CLAMP

BELT ASSEMBLY

DATA LIST POCKET

URINE TRANSFER
CONNECTOR AND
OMEDICAL INJECTION
FLAP

ASSIST STRAP

SLIDE FASTENER

BOOT

LOOP TAPE

LOOP TAPE

LM REST

ENTRANCE
SLIDE FASTENER
FLAP

ACTIVE
DOSIMETER
POCKET

LANYARD POCKET

ASSISTS

SCISSORS POCKET

CHECKLIST POCKET

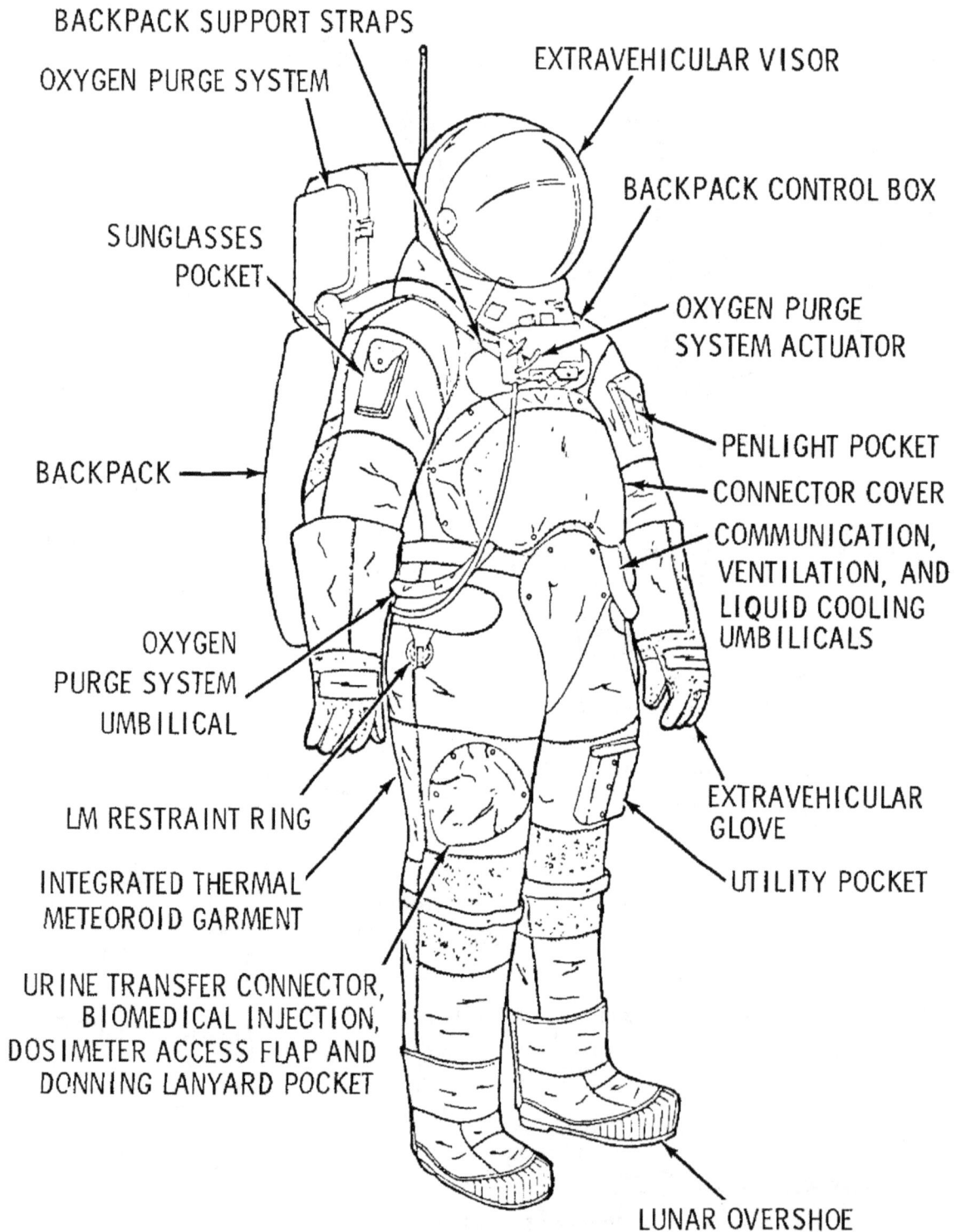

BACKPACK SUPPORT STRAPS

OXYGEN PURGE SYSTEM

EXTRAVEHICULAR VISOR

SUNGLASSES POCKET

BACKPACK CONTROL BOX

OXYGEN PURGE SYSTEM ACTUATOR

BACKPACK

PENLIGHT POCKET

CONNECTOR COVER

COMMUNICATION, VENTILATION, AND LIQUID COOLING UMBILICALS

OXYGEN PURGE SYSTEM UMBILICAL

EXTRAVEHICULAR GLOVE

LM RESTRAINT RING

UTILITY POCKET

INTEGRATED THERMAL METEOROID GARMENT

URINE TRANSFER CONNECTOR, BIOMEDICAL INJECTION, DOSIMETER ACCESS FLAP AND DONNING LANYARD POCKET

LUNAR OVERSHOE

Extravehicular visor assembly--A polycarbonate shell and two visors with thermal control and optical coatings on them. The EVA visor is attached over the pressure helmet to provide impact, micrometeoroid, thermal and light protection to the EVA crewman.

Extravehicular gloves--Built of an outer shell of Chromel-R fabric and thermal insulation to provide protection when handling extremely hot and cold objects. The finger tips are made of silicone rubber to provide the crewman more sensitivity.

A one-piece constant-wear garment, similar to "long johns", is worn as an undergarment for the space suit in intravehicular operations and for the inflight coveralls. The garment is porous-knit cotton with a waist-to-neck zipper for donning. Biomedical harness attach points are provided.

During periods out of the space suits, crewmen will wear two-piece Teflon fabric inflight coveralls for warmth and for pocket stowage of personal items.

Communications carriers ("Snoopy hats") with redundant microphones and earphones are worn with the pressure helmet; a lightweight headset is worn with the inflight coveralls.

Meals

The Apollo 10 crew has a wide range of food items from which to select their daily mission space menu. More than 60 items comprise the food selection list of freeze-dried rehydratable foods. In addition, one "wet pack" meal-per-man per-day will be stowed for a total of 27. These meals, consisting of foil-wrapped beef and potatoes, ham and potatoes and turkey chunks and gravy, are similar to the Christmas meals carried aboard Apollo 8 and can be eaten with a spoon.

Water for drinking and rehydrating food is obtained from three sources in the command module -- a dispenser for drinking water and two water spigots at the food preparation station, one supplying water at about 155 degrees F., the other at about 55 degrees F. The potable water dispenser squirts water continuously as long as the trigger is held down, and the food preparation spigots dispense water in one-ounce increments.

Command module potable water is supplied from service module fuel cell byproduct water. Three one-pint "picnic jugs", or plastic bags, will be stowed aboard Apollo 10 for drinking water. Each crewman once a day will fill a bag with water and then spin it up to separate the suspended hydrogen gas from the water so that he will have hydrogen-less water to drink the following day. The suspended hydrogen in the fuel cell byproduct water has caused intestinal dis-comfort to crewmen in previous Apollo missions.

A continuous-feed hand water dispenser similar to the one in the command module is used aboard the lunar module for cold-water rehydration of food packets stowed aboard the LM.

After water has been injected into a food bag, it is kneaded for about three minutes. The bag neck is then cut off and the food squeezed into the crewman's mouth. After a meal, germicide pills attached to the outside of the food bags are placed in the bags to prevent fermentation and gas formation. The bags are then rolled and stowed in waste disposal compart-ments.

The day-by-day, meal-by-meal Apollo 10 menu for each crew-man for both the command module and the lunar module is listed on the following pages.

APOLLO 10 (STAFFORD)

MEAL	Day 1*, 5, 9	Day 2, 6, 10	Day 3, 7, 11	Day 4, 8
A	Peaches Bacon Squares (8) Cinn Tstd Bread Cubes (4) Grapefruit Drink Orange Drink	Fruit Cocktail Sugar Coated Corn Flakes Bacon Squares (8) Grapefruit Drink Grape Drink	Peaches Bacon Squares (8) Strawberry Cubes (4) Cocoa Orange Drink	Fruit Cocktail Sausage Patties Bacon Squares (8) Cocoa Grape Drink
B	Salmon Salad Chicken & Rice** Sugar Cookie Cubes (4) Cocoa Grape Punch	Potato Soup Chicken & Vegetables Tuna Salad Pineapple Fruitcake (4) Orange Drink	Cream of Chicken Soup (Turkey & Gravy - Wet Pack) Butterscotch Pudding Brownies (4) Grapefruit Drink	Potato Soup Pork & Scalloped Potatoes Applesauce Orange Drink
C	(Beef & Potatoes - Wet Pack) Cheese Cracker Cubes (4) Chocolate Pudding Orange-Grapefruit Drink	Spaghetti & Meat Sauce** (Ham & Potatoes - Wet Pack) Banana Pudding Pineapple-Grapefruit Drink	Pea Soup Beef Stew** Chicken Salad Chocolate Cubes (4) Grape Punch	Shrimp Cocktail Chicken Stew** Turkey Bites (4) Date Fruitcake (4) Orange-Grapefruit Drink
CALORIES/DAY	2172	2179	2530	201

*Day 1 consists of Meal C only
**New spoon-bowl package

-more-

APOLLO 10 (YOUNG)

MEAL	Day 1*, 5, 9	Day 2, 6, 10	Day 3, 7, 11	Day 4, 8
A	Peaches Bacon Squares (8) Cinn Tstd Bread Cubes (4) Grapefruit Drink Orange Drink	Fruit Cocktail Sugar Coated Corn Flakes Brownies (4) Grapefruit Drink Grape Drink	Peaches Bacon Squares (8) Strawberry Cubes (4) Cocoa Orange Drink	Fruit Cocktail Sausage Patties Bacon Squares (8) Cocoa Grape Drink
B	Salmon Salad Chicken & Rice** Sugar Cookie Cubes (4) Cocoa Grape Punch	Potato Soup Tuna Salad Chicken & Vegetables Pineapple Fruitcake (4) Pineapple-Grapefruit Drink	Cream of Chicken Soup (Turkey & Gravy - Wet Pack) Butterscotch Pudding Grapefruit Drink	Pea Soup Pork & Scalloped Potatoes Applesauce Orange Drink
C	(Beef & Potatoes - Wet Pack) Cheese Cracker Cubes (4) Chocolate Pudding Orange-Grapefruit Drink	Spaghetti & Meat Sauce** (Ham & Potatoes - Wet Pack) Banana Pudding Orange Drink	Beef Stew** Chicken Salad Corn Chowder Chocolate Cubes (4) Grape Punch	Shrimp Cocktail Chicken Stew** Turkey Bites (4) Date Fruitcake (4) Orange-Grapefruit Drink
CALORIES/DAY	2172	2145	2389	X

*Day 1 consists of Meal C only
**New spoon-bowl package

APOLLO 10 (CERNAN)

MEAL	Day 1*, 5, 9	Day 2, 6, 10	Day 3, 7, 11	Day 4, 8
A	Peaches Bacon Squares (8) Cinn Tstd Bread Cubes (4) Orange Drink Orange-Pineapple Drink	Fruit Cocktail Sugar Coated Corn Flakes Bacon Squares (8) Orange Drink Grape Drink	Peaches Bacon Squares (8) Strawberry Cubes (4) Cocoa Orange Drink	Fruit Cocktail Sausage Patties Bacon Squares (8) Cocoa Grape Drink
B	Salmon Salad Chicken & Rice** Sugar Cookie Cubes (4) Cocoa Grape Punch	Potato Soup Tuna Salad Chicken & Vegetables Brownies (4) Orange-Grapefruit Drink	Cream of Chicken Soup (Turkey & Gravy - Wet Pack) Cinn Tstd Bread Cubes (4) Butterscotch Pudding Pineapple-Grapefruit Drink	Potato Soup Pork & Scalloped Potatoes Applesauce Orange Drink
C	Cream of Chicken Soup (Beef & Potatoes - Wet Pack) Cheese Cracker Cubes (4) Fruit Cocktail Orange-Grapefruit Drink	Spaghetti & Meat Sauce** (Ham & Potatoes - Wet Pack) Banana Pudding Orange Drink	Pea Soup Chicken Salad Beef Stew** Grape Punch	Shrimp Cocktail Chicken Stew** Turkey Bites (6) Chocolate Cubes (6) Orange-Grapefruit Drink
CALORIES/DAY	2026	2040	2298	2021

*Day 1 consists of Meal C only
**New spoon-bowl package

-more-

APOLLO 10 LM MENU

Day 1

Meal A

Fruit Cocktail
Bacon Squares (8)
Brownies (4)
Orange Drink
Grape Punch

Meal B

Beef and Vegetables
Pineapple Fruitcake (4)
Orange-Grapefruit Drink
Grape Punch

Meal C

Cream of Chicken Soup
Beef Hash
Strawberry Cubes (4)
Pineapple-Grapefruit Drink

2 man-days only
2 meals per overwrap
Red and Blue Velcro

Personal Hygiene

Crew personal hygiene equipment aboard Apollo 10 includes body cleanliness items, the waste management system and one medical kit.

Packaged with the food are a toothbrush and a two-ounce tube of toothpaste for each crewman. Each man-meal package contains a 3.5-by-four-inch wet-wipe cleansing towel. Additionally, three packages of 12-by-12-inch dry towels are stowed beneath the command module pilot's couch. Each package contains seven towels. Also stowed under the command module pilot's couch are seven tissue dispensers containing 53 three-ply tissues each.

Solid body wastes are collected in Gemini-type plastic defecation bags which contain a germicide to prevent bacteria and gas formation. The bags are sealed after use and stowed in empty food containers for post-flight analysis.

Urine collection devices are provided for use while wearing either the pressure suit or the inflight coveralls. The urine is dumped overboard through the spacecraft urine dump valve in the CM and stored in the LM.

The 5x5x8-inch medical accessory kit is stowed in a compartment on the spacecraft right side wall beside the lunar module pilot couch. The medical kit contains three motion sickness injectors, three pain suppression injectors, one two-ounce bottle first aid ointment, two one-ounce bottle eye drops, three nasal sprays, two compress bandages, 12 adhesive bandages, one oral thermometer and two spare crew biomedical harnesses. Pills in the medical kit are 60 antibiotic, 12 nausea, 12 stimulant, 18 pain killer, 60 decongestant, 24 diarrhea, 72 aspirin and 21 sleeping. Additionally, a small medical kit containing four stimulant, eight diarrhea, two sleeping and four pain killer pills, 12 aspirin, one bottle eye drops and two compress bandages is stowed in the lunar module flight data file compartment.

Survival Gear

The survival kit is stowed in two rucksacks in the right-hand forward equipment bay above the lunar module pilot.

Contents of rucksack No. 1 are: two combination survival lights, one desalter kit, three pair sunglasses, one radio beacon, one spare radio beacon battery and spacecraft connector cable, one knife in sheath, three water containers and two containers of Sun lotion.

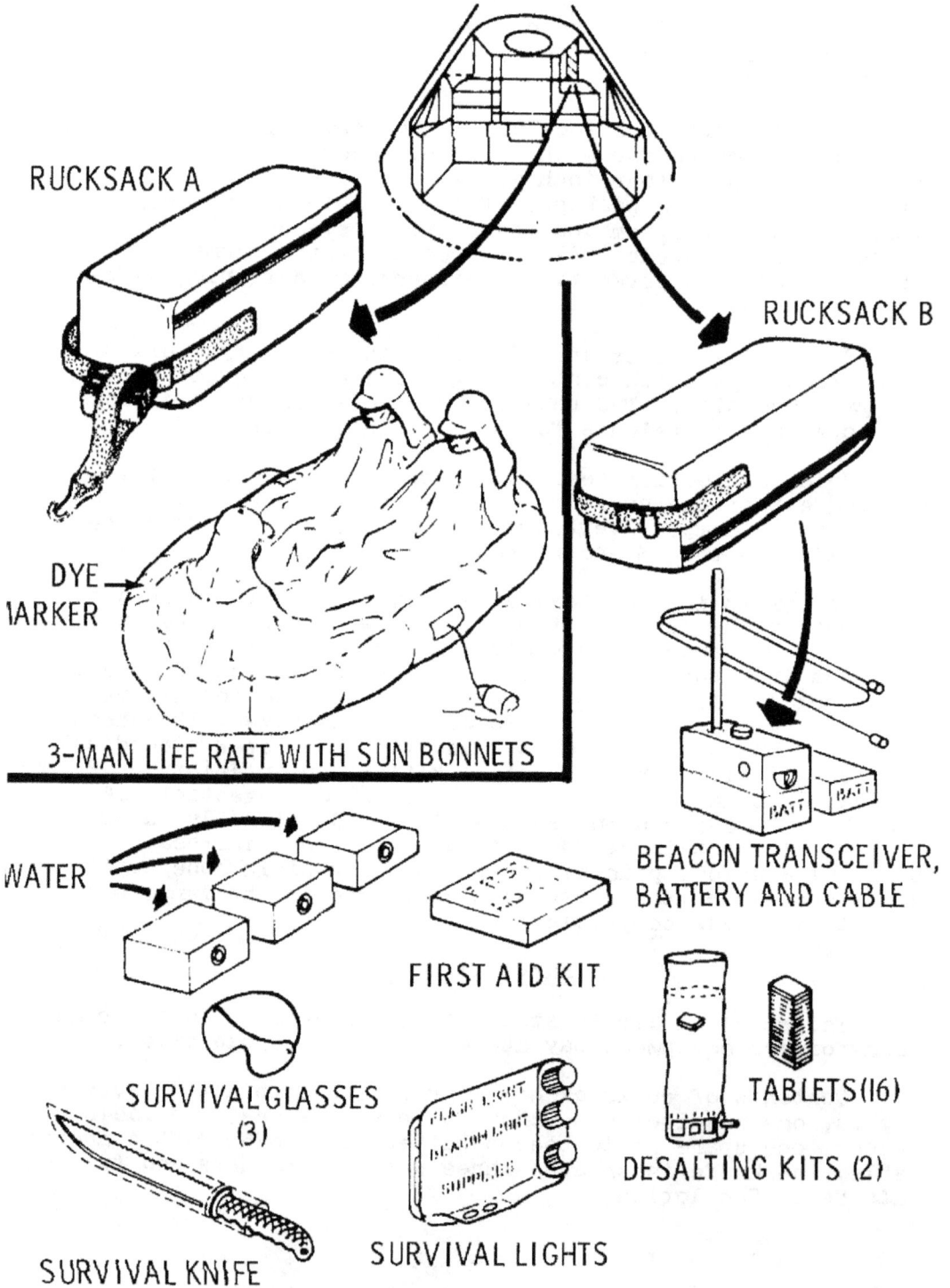

RUCKSACK A

RUCKSACK B

DYE MARKER

3-MAN LIFE RAFT WITH SUN BONNETS

BEACON TRANSCEIVER, BATTERY AND CABLE

WATER

FIRST AID KIT

SURVIVAL GLASSES (3)

TABLETS (16)

DESALTING KITS (2)

SURVIVAL KNIFE

SURVIVAL LIGHTS

Rucksack No. 2: one three-man life raft with CO_2 inflater, one sea anchor, two sea dye markers, three sun-bonnets, one mooring lanyard, three manlines, and two attach brackets.

The survival kit is designed to provide a 48-hour postlanding (water or land) survival capability for three crewmen between 40 degrees North and South latitudes.

Biomedical Inflight Monitoring

The Apollo 10 crew biomedical telemetry data received by the Manned Space Flight Network will be relayed for in-stantaneous display at Mission Control Center where heart rate and breathing rate data will be displayed on the flight surgeon's console. Heart rate and respiration rate average, range and deviation are computed and displayed on digital TV screens.

In addition, the instantaneous heart rate, real-time and delayed EKG and respiration are recorded on strip charts for each man.

Biomedical telemetry will be simultaneous from all crew-men while in the CSM, but selectable by a manual onboard switch in the LM.

Biomedical data observed by the flight surgeon and his team in the Life Support Systems Staff Support Room will be correlated with spacecraft and space suit environmental data displays.

Blood pressures are no longer telemetered as they were in the Mercury and Gemini programs. Oral temperature, how-ever, can be measured onboard for diagnostic purposes and voiced down by the crew in case of inflight illness.

Rest-Work Cycles

All three Apollo 10 crewmen will sleep simultaneously during rest periods. The average mission day will consist of 16 hours of work and eight hours of rest. Two crewmen normally will sleep in the sleep stations (sleeping bags) under the couches, with the third man in the couch. During rest periods, one crewman will wear his communications headset.

The only exception to this sleeping arrangement will be during the rest period on lunar orbit insertion day, when two crewmen will sleep in the couches since the docking probe and drogue assemblies will be stowed in one of the sleep stations.

When possible, all three crewmen will eat together in one-hour eat periods during which other activities will be held to a minimum.

more

Training

The crewmen of Apollo 10 have spent more than five hours of formal crew training for each hour of the lunar-orbit mission's eight-day duration. Almost 1,000 hours of training were in the Apollo 10 crew training syllabus over and above the normal preparations for the mission--technical briefings and reviews, pilot meetings and study.

The Apollo 10 crewmen also took part in spacecraft manu-facturing checkouts at the North American Rockwell plant in Downey, Calif., at Grumman Aircraft Engineering Corp., Bethpage, N.Y., and in prelaunch testing at NASA Kennedy Space Center. Taking part in factory and launch area testing has provided the crew with thorough operational knowledge of the complex vehicle.

Highlights of specialized Apollo 10 crew training topics are:

* Detailed series of briefings on spacecraft systems, operation and modifications.

* Saturn launch vehicle briefings on countdown, range safety, flight dynamics, failure modes and abort conditions. The launch vehicle briefings were updated periodcally.

* Apollo Guidance and Navigation system briefings at the Massachusetts Institute of Technology Instrumentation Laboratory.

* Briefings and continuous training on mission photo-graphic objectives and use of camera equipment.

* Extensive pilot participation in reviews of all flight procedures for normal as well as emergency situations.

* Stowage reviews and practice in training sessions in the spacecraft, mockups and command module simulators allowed the crewmen to evaluate spacecraft stowage of crew-associated equipment.

* More than 300 hours of training per man in command module and lunar module simulators at MSC and KSC, including closed-loop simulations with flight controllers in the Mission Control Center. Other Apollo simulators at various locations were used extensively for specialized crew training.

* Entry corridor deceleration profiles at lunar-return conditions in the MSC Flight Acceleration Facility manned centrifuge.

     * Zero-g aircraft flights using command module and lunar
module mockups for EVA and pressure suit doffing/donning
practice and training.

     * Underwater zero-g training in the MSC Water Immersion
Facility using spacecraft mockups to familiarize further crew
with all aspects of CSM-LM docking tunnel intravehicular
transfer and EVA in pressurized suits.

     * Water egress training conducted in indoor tanks as
well as in the Gulf of Mexico, included uprighting from the
Stable II position (apex down) to the Stable I position
(apex up), egress onto rafts and helicopter pickup.

     * Launch pad egress training from mockups and from the
actual spacecraft on the launch pad for possible emergencies
such as fire, contaminants and power failures.

     * The training covered use of Apollo spacecraft fire
suppression equipment in the cockpit.

     * Planetarium reviews at Morehead Planetarium, Chapel
Hill, N.C., and at Griffith Planetarium, Los Angeles, Calif.,
of the celestial sphere with special emphasis on the 37
navigational stars used by the Apollo guidance computer.

Crew Biographies

NAME: Thomas P. Stafford (Colonel, USAF) Apollo 10 commander
NASA Astronaut

BIRTHPLACE AND DATE: Born September 17, 1930, in Weatherford,
Okla., where his mother, Mrs. Mary Ellen Stafford, now
resides.

PHYSICAL DESCRIPTION: Black hair, blue eyes; height: 6 feet;
weight: 175 pounds.

EDUCATION: Graudated from Weatherford High School, Weatherford,
Okla.; received a Bachelor of Science degree from the
United States Naval Academy in 1952; recipient of an
Honorary Doctorate of Science from Oklahoma City University
in 1967.

MARITAL STATUS: Married to the former Faye L. Shoemaker of
Weatherford, Okla. Her parents, Mr. and Mrs. Earle R.
Shoemaker, reside in Thomas, Okla.

CHILDREN: Dionne, July 2, 1954; Karin, Aug. 28, 1957.

OTHER ACTIVITIES: His hobbies include handball, weight lifting
and swimming.

ORGANIZATIONS: Member of the Society of Experimental Test
Pilots.

SPECIAL HONORS: Awarded two NASA Exceptional Service Medals
and the Air Force Astronaut Wings; the Distinguished
Flying Cross; the AIAA Astronautics Award; and co-re-
cipient of the 1966 Harmon International Aviation Trophy.

EXPERIENCE: Stafford, an Air Force colonel, was commissioned in
in the United States Air Force upon graduation from
Annapolis. Following his flight training, he flew fighter
interceptor aircraft in the United States and Germany
and later attended the USAF Experimental Flight Test
School at Edwards Air Force Base, Calif.

He served as Chief of the Performance Branch at the USAF
Aerospace Research Pilot School at Edwards and was responsible for the supervision and administration of the
flying curriculum for student test pilots. He was also
an instructor in flight test training and specialized
academic subjects--establishing basic textbooks and
directing the writing of flight test manuals for use by
the staff and students. He is co-author of the Pilot's
Handbook for Performance Flight Testing and the Aero-
dynamics Handbook for Performance Flight Testing.

-more-

He has accumulated over 5,000 hours flying time, of which over 4,000 hours are in jet aircraft.

CURRENT ASSIGNMENT: Colonel Stafford was selected as an astronaut by NASA in September 1962. He has since served as backup pilot for the Gemini 3 flight.

On Dec. 15, 1965, he and command pilot Walter M. Schirra were launched into space on the history-making Gemini 6 mission and subsequently participated in the first successful rendezvous of two manned maneuverable space-craft by joining the already orbiting Gemini 7 crew. Gemini 6 returned to Earth on Dec. 16, 1965, after 25 hours 51 minutes and 24 seconds of flight.

He made his second flight as command pilot of the Gemini 9 mission. During this 3-day flight which began on June 3, 1966, the spacecraft attained a circular orbit of 161 statute miles; the crew performed three different types of rendezvous with the previously launched Augmented Target Docking Adapter; and pilot Eugene Cernan logged two hours and ten minutes outside the spacecraft in extravehicular activity. The flight ended after 72 hours and 20 minutes with a perfect reentry and recovery as Gemini 9 landed within 0.4 nautical miles of the de-signated target point and 1½ miles from the prime recovery ship, USS WASP.

NAME: John W. Young(Commander, USN) Apollo 10 command module
        pilot
        NASA Astronaut

BIRTHPLACE AND DATE: Born in San Francisco, Calif., on Sept.
        24, 1930. His parents, Mr. and Mrs. William H. Young,
        reside in Orlando, Fla.

PHYSICAL DESCRIPTION: Brown hair; green eyes; height: 5 feet
        9 inches; weight: 165 pounds.

EDUCATION: Graudated from Orlando High School, Orlando, Fla.;
        received a Bachelor of Science degree in Aeronautical
        Engineering from the Georgia Institute of Technology in
        1952.

MARITAL STATUS: Married to the former Barbara V. White of
        Savannah, Ga. Her parents, Mr. and Mrs. Robert A. White,
        reside in Jacksonville, Fla.

CHILDREN: Sandy, Apr. 30, 1957; John, Jan. 17, 1959.

OTHER ACTIVITIES: His hobbies are bicycle riding and handball.

ORGANIZATIONS: Member of the American Institute of Aeronautics
        and Astronautics and the Society of Experimental Test
        Pilots.

SPECIAL HONORS: Awarded two NASA Exceptional Service Medals,
        the Navy Astronaut Wings, and three Distinguished Flying
        Crosses.

EXPERIENCE: Upon graduation from Georgia Tech, Young entered
        the U.S. Navy in 1952 and holds the rank of commander.

        He was a test pilot at the Naval Air Test Center from 1959
        to 1962. Test projects included evaluations of the F8D
        and F4B fighter weapons systems. In 1962, he set world
        time-to-climb records to 3,000 and 25,000-meter altitudes
        in the F4B. Prior to his assignment to NASA he was
        Maintenance Officer of All-Weather-Fighter Squadron 143
        at the Naval Air Station, Miramar, Calif.

        He has logged more than 4,500 hours flying time, including
        more than 3,900 hours in jet aircraft.

CURRENT ASSIGNMENT: Commander Young was selected as an astro-
     naut by NASA in September 1962.

He served as pilot on the first manned Gemini flight--a
3-orbit mission, launched on March 23, 1965, during which
the crew accomplished the first manned spacecraft orbital
trajectory modifications and lifting reentry, and flight
tested all systems in Gemini 3. After this assignment, he
was backup pilot for Gemini 6.

On July 18, 1966, Young occupied the command pilot seat
for the Gemini 10 mission and, with Michael Collins as
pilot, effected a successful rendezvous and docking with
the Agena target vehicle. Then, they ignited the large
Agena main engine to propel the docked combination to
a record altitude of approximately 475 miles above the
Earth--the first manned operation of a large rocket
engine in space. They later performed a completely
optical rendezvous (without radar) on a second passive
Agena. After the rendezvous, while Young flew formation
on the passive Agena, Collins performed extravehicular
activity to it and recovered a micrometeorite detection
experiment, accomplishing an in-space retrieval of the
detector that had been orbiting the Earth for three months.

The flight was concluded after 3 days and 44 revolutions--
during which Gemini 10 traveled a total distance of 1,275,
091 statute miles. Splashdown occurred in the West Atlantic,
529 statute miles east of Cape Kennedy, where Gemini 10
landed 2.6 miles from the USS GUADALCANAL within eye and
camera range of the prime recovery vessel.

-more-

NAME: Eugene A. Cernan (Commander, USN) Apollo 10 lunar module
   pilot
   NASA Astronaut

BIRTHPLACE AND DATE: Born in Chicago, Ill., on March 14, 1934.
   His mother, Mrs. Andrew G. Cernan, resides in Bellwood,
   Ill.

PHYSICAL DESCRIPTION: Brown hair; blue eyes; height: 6 feet;
   weight: 170 pounds.

EDUCATION: Graduated from Proviso Township High School in
   Maywood, Ill.; received a Bachelor of Science degree in
   Electrical Engineering from Purdue University and a Master
   of Science degree in Aeronautical Engineering from the
   U.S. Naval Postgraduate School.

MARITAL STATUS: Married to the former Barbara J. Atchley of
   Houston, Tex.

CHILDREN: Teresa Dawn, March 4, 1963.

OTHER ACTIVITIES: His hobbies include gardening and all sports
   activities.

ORGANIZATIONS: Member of Tau Beta Pi, national engineering society;
   Sigma Xi, national science research society; and Phi Gamma
   Delta, national social fraternity.

SPECIAL HONORS: Awarded the NASA Exceptional Service Medal; the
   Navy Astronaut Wings; and the Distinguished Flying Cross.

EXPERIENCE: Cernan, a United States Navy commander, received his
   commission through the Navy ROTC program at Purdue. He
   entered flight training upon his graduation.

   Prior to attending the Naval Postgraduate School, he was
   assigned to Attack Squadrons 126 and 113 at the Miramar,
   Calif., Naval Air Station.

   He has logged more than 3,000 hours flying time with more
   than 2,810 hours in jet aircraft.

CURRENT ASSIGNMENT: Commander Cernan was one of the third group
   of astronauts selected by NASA in October 1963.

He occupied the pilot seat alongside Command Pilot Tom
Stafford on the Gemini 9 mission. During this 3-day
flight which began on June 3, 1966, the spacecraft attained
a circular orbit of 161 statute miles; the crew used three
different techniques to effect rendezvous with the previously launched Augmented Target Docking Adapter; and
Cernan logged two hours and ten minutes outside the spacecraft in extravehicular activity. The flight ended after
72 hours and 20 minutes with a perfect reentry and recovery as Gemini 9 landed within 1½ miles of the prime
recovery ship USS WASP and 3/8 of a mile from the pre-
determined target point.

He has since served as backup pilot for Gemini 12.

APOLLO LAUNCH OPERATIONS

Prelaunch Preparations

NASA's John F. Kennedy Space Center performs preflight checkout, test, and launch of the Apollo 10 space vehicle. A government-industry team of about 550 will conduct the final countdown from Firing Room 3 of the Launch Control Center (LCC).

The firing room team is backed up by more than 5,000 persons who are directly involved in launch operations at KSC -- from the time the vehicle and spacecraft stages arrive at the center until the launch is completed.

Initial checkout of the Apollo spacecraft is conducted in work stands and in the altitude chambers in the Manned Space-craft Operations Building (MSOB) at Kennedy Space Center. After completion of checkout there, the assembled spacecraft is taken to theVehicle Assembly Building (VAB) and mated with the launch vehicle. There the first integrated spacecraft and launch vehicle tests are conducted. The assembled space vehicle is then rolled out to the luanch pad for final preparations and countdown to launch.

In mid-October 1968, flight hardware for Apollo 10 began arriving at Kennedy Space Center, just as Apollo 7 was being launched from Complex 34 on Cape Kennedy and as Apollo 8 and Apollo 9 were undergoing checkout at Kennedy Space Center.

The lunar module was the first piece of Apollo 10 flight hardware to arrive at KSC. The two stages were moved into the altitude chamber in the Manned Spacecraft Operations Building (MSOB) after an initial receiving inspection in October. In the chamber the LM underwent systems tests and both unmanned and manned chamber runs. During these runs the chamber air was pumped out to simulate the vacuum of space at altitudes in excess of 200,000 feet. There the spacecraft systems and the astronauts' life support systems were tested.

While the LM was undergoing preparation for its manned altitude chamber runs, the Apollo 10 command/service module arrived at KSC and after receiving inspection, it, too, was placed in an altitude chamber in the MSOB for systems tests and unmanned and manned chamber runs. The prime and back-up crews participated in the chamber runs on both the LM and the CSM.

In January, the LM and CSM were removed from the chambers. After installing the landing gear on the LM and the SPS engine nozzle on the CSM, the LM was encapsulated in the spacecraft LM adapter (SLA) and the CSM was mated to the SLA. On February 6, the assembled spacecraft was moved to the VAB where it was mated to the launch vehicle.

The launch vehicle flight hardware began arriving at KSC in late November, and by the end of December the three stages and the instrument unit were erected on the mobile launcher in high bay 2. This was the first time high bay 2, on the west side of the VAB, had been used for assembling a Saturn V. Tests were conducted on individual systems on each of the stages and on the overall launch vehicle before the spacecraft was erected atop the vehicle.

After spacecraft erection, the spacecraft and launch vehicle were electrically mated and the first overall test (plugs-in) of the space vehicle was conducted. In accordance with the philosophy of accomplishing as much of the checkout as possible in the VAB, the overall test was conducted before the space vehicle was moved to the launch pad.

The plugs-in test verified the compatibility of the space vehicle systems, ground support equipment, and off-site support facilities by demonstrating the ability of the systems to proceed through a simulated countdown, launch, and flight. During the simulated flight portion of the test, the systems were required to respond to both emergency and normal flight conditions.

The move to Pad B from the VAB on March 11 occurred while the Apollo 9 circled the Earth in the first manned test of the lunar module.

Apollo 10 will mark the first launch at Pad B on complex 39. The first two unmanned Saturn V launches and the manned Apollo 8 and 9 launches took place at Pad A. It also marked the first time that the transporter maneuvered around the VAB carrying a full load from high bay 2 on the 5-mile trip to the launch pad.

The space vehicle Flight Readiness Test was conducted in early April. Both the prime and backup crews participate in portions of the FRT, which is a final overall test of the space vehicle systems and ground support equipment when all systems are as near as possible to a launch configuration.

After hypergolic fuels were loaded aboard the space vehicle, and the launch vehicle first stage fuel (RP-1) was brought aboard, the final major test of the space vehicle began. This was the countdown demonstration test (CDDT), a dress rehearsal for the final countdown to launch. The CDDT for Apollo 10 was divided into a "wet" and a "dry" portion. During the first, or "wet" portion, the entire countdown, including propellant loading, was carried out down to T-8.9 seconds. The astronaut crews did not participate in the wet CDDT. At the completion of the wet CDDT, the cryogenic propellants (liquid oxygen and liquid hydrogen) were off-loaded, and the final portion of the countdown was re-run, this time simulating the fueling and with the prime astronaut crew participating as they will on launch day.

By the time Apollo 10 was entering the final phase of its checkout procedure at Complex 39B, crews had already started the checkout of Apollo 11 and Apollo 12. The Apollo 11 spacecraft completed altitude chamber testing and was mated to the launch vehicle in the VAB in mid-April as the Apollo 12 CSM and LM began checkout in the altitude chambers.

Because of the complexity involved in the checkout of the 363-foot-tall (110.6 meters) Apollo/Saturn V configuration, the launch teams make use of extensive automation in their checkout. Automation is one of the major differences in checkout used on Apollo compared to the procedures used in the Mercury and Gemini programs.

Computers, data display equipment, and digital data techniques are used throughout the automatic checkout from the time the launch vehicle is erected in the VAB through liftoff. A similar, but separate computer operation called ACE (Acceptance Checkout Equipment) is used to verify the flight readiness of the spacecraft. Spacecraft checkout is controlled from separate rooms in the Manned Spacecraft Operations Building.

LAUNCH COMPLEX 39

Launch Complex 39 facilities at the Kennedy Space Center were planned and built specifically for the Apollo Saturn V program, the space vehicle that will be used to carry astronauts to the Moon.

Complex 39 introduced the mobile concept of launch operations, a departure from the fixed launch pad techniques used previously at Cape Kennedy and other launch sites. Since the early 1950's when the first ballistic missiles were launched, the fixed launch concept had been used on NASA missions. This method called for assembly, checkout and launch of a rocket at one site--the launch pad. In addition to tying up the pad, this method also often left the flight equipment exposed to the outside influences of the weather for extended periods.

Using the mobile concept, the space vehicle is thoroughly checked in an enclosed building before it is moved to the launch pad for final preparations. This affords greater protection, a more systematic checkout process using computer techniques and a high launch rate for the future, since the pad time is minimal.

Saturn V stages are shipped to the Kennedy Space Center by ocean-going vessels and specially designed aircraft, such as the Guppy. Apollo spacecraft modules are transported by air. The spacecraft components are first taken to the Manned Spacecraft Operations Building for preliminary checkout. The Saturn V stages are brought immediately to the Vehicle Assembly Building after arrival at the nearby turning basin.

Apollo 10 is the first vehicle to be launched from Pad B, Complex 39. All previous Saturn V vehicles were launched Pad A at Complex 39. The historic first launch of the Saturn V, designated Apollo 4, took place Nov. 9, 1967 after a perfect countdown and on-time liftoff at 7 a.m. EST. The second Saturn V mission--Apollo 6--was conducted last April 4. The third Saturn V mission, Apollo 8, was conducted last Dec. 21-27. Apollo 9 was March 3-13, 1969.

The major components of Complex 39 include: (1) the Vehicle Assembly Building (VAB) where the Apollo 10 was assembled and prepared; (2) the Launch Control Center, where the launch team conducts the preliminary checkout and final countdown; (3) the mobile launcher, upon which the Apollo 10 was erected for checkout and from where it will be launched; (4) the mobile service structure, which provides external access to the space vehicle at the pad; (5) the transporter, which arries the space vehicle and mobile launcher, as well as the mobile service structure to the pad; (6) the crawlerway over which the space vehicle travels from the VAB to the launch pad; and (7) the launch pad itselft.

-more-

Vehicle Assembly Building

The Vehicle Assembly Building is the heart of Launch Complex 39. Covering eight acres, it is where the 363-foot-tall space vehicle is assembled and tested.

The VAB contains 129,482,000 cubic feet of space. It is 716 feet long, and 518 feet wide and it covers 343,500 square feet of floor space.

The foundation of the VAB rests on 4,225 steel pilings, each 16 inches in diameter, driven from 150 to 170 feet to bedrock. If placed end to end, these pilings would extend a distance of 123 miles. The skeletal structure of the building contains approximately 60,000 tons of structural steel. The exterior is covered by more than a million square feet of insulated aluminum siding.

The building is divided into a high bay area 525 feet high and a low bay area 210 feet high, with both areas serviced by a transfer aisle for movement of vehicle stages.

The low bay work area, approximately 442 feet wide and 274 feet long, contains eight stage-preparation and checkout cells. These cells are equipped with systems to simulate stage interface and operation with other stages and the instrument unit of the Saturn V launch vehicle.

After the Apollo 10 launch vehicle upper stages arrived at the Kennedy Space Center, they were moved to the low bay of the VAB. Here, the second and third stages underwent acceptance and checkout testing prior to mating with the S-IC first stage atop mobile launcher 3 in the high bay area.

The high bay provides facilities for assembly and checkout of both the launch vehicle and spacecraft. It contains four separate bays for vertical assembly and checkout. At present, three bays are equipped, and the fourth will be reserved for possible changes in vehicle configuration.

Work platforms -- some as high as three-story buildings -- in the high bays provide access by surrounding the vehicle at varying levels. Each high bay has five platforms. Each platform consists of two bi-parting sections that move in from opposite sides and mate, providing a 360-degree access to the section of the space vehicle being checked.

A 10,000-ton-capacity air conditioning system, sufficient to cool about 3,000 homes, helps to control the environment within the entire office, laboratory, and workshop complex located inside the low bay area of the VAB. Air conditioning is also fed to individual platform levels located around the vehicle.

-more-

There are 141 lifting devices in the VAB, ranging from one-ton hoists to two 250-ton high-lift bridge cranes.

The mobile launchers, carried by transporter vehicles, move in and out of the VAB through four doors in the high bay area, one in each of the bays. Each door is shaped like an inverted T. They are 152 feet wide and 114 feet high at the base, narrowing to 76 feet in width. Total door height is 456 feet.

The lower section of each door is of the aircraft hangar type that slides horizontally on tracks. Above this are seven telescoping vertical lift panels stacked one above the other, each 50 feet high and driven by an individual motor. Each panel slides over the next to create an opening large enough to permit passage of the mobile launcher.

Launch Control Center

Adjacent to the VAB is the Launch Control Center (LCC). This four-story structure is a radical departure from the dome-shaped blockhouses at other launch sites.

The electronic "brain" of Launch Complex 39, the LCC was used for checkout and test operations while Apollo 10 was being assembled inside the VAB. The LCC contains display, monitoring, and control equipment used for both checkout and launch operations.

The building has telemeter checkout stations on its second floor, and four firing rooms, one for each high bay of the VAB, on its third floor. Three firing rooms contain identical sets of control and monitoring equipment, so that launch of a vehicle and checkout of others take place simultaneously. A ground computer facility is associated with each firing room.

The high speed computer data link is provided between the LCC and the mobile launcher for checkout of the launch vehicle. This link can be connected to the mobile launcher at either the VAB or at the pad.

The three equipped firing rooms have some 450 consoles which contain controls and displays required for the checkout process. The digital data links connecting with the high bay areas of the VAB and the launch pads carry vast amounts of data required during checkout and launch.

There are 15 display systems in each LCC firing room, with each system capable of providing digital information instantaneously.

Sixty television cameras are positioned around the Apollo/ Saturn V transmitting pictures on 10 modulated channels. The LCC firing room also contains 112 operational intercommunication channels used by the crews in the checkout and launch countdown.

Mobile Launcher

The mobile launcher is a transportable launch base and umbilical tower for the space vehicle. Three mobile launchers are used at Complex 39.

The launcher base is a two-story steel structure, 25 feet high, 160 feet long, and 135 feet wide. It is positioned on six steel pedestals 22 feet high when in the VAB or at the launch pad. At the launch pad, in addition to the six steel pedestals, four extendable columns also are used to stiffen the mobile launcher against rebound loads, if the Saturn engines cut off.

The umbilical tower, extending 398 feet above the launch platform, is mounted on one end of the launcher base. A hammerhead crane at the top has a hook height of 376 feet above the deck with a traverse radius of 85 feet from the center of the tower.

The 12-million-pound mobile launcher stands 445 feet high when resting on its pedestals. The base, covering about half an acre, is a compartmented structure built of 25-foot steel girders.

The launch vehicle sits over a 45-foot-square opening which allows an outlet for engine exhausts into the launch pad trench containing a flame deflector. This opening is lined with a replaceable steel blast shield, independent of the structure, and is cooled by a water curtain initiated two seconds after liftoff.

There are nine hydraulically-operated service arms on the umbilical tower. These service arms support lines for the vehicle umbilical systems and provide access for personnel to the stages as well as the astronaut crew to the spacecraft.

On Apollo 10, one of the service arms is retracted early in the count. The Apollo spacecraft access arm is partially retracted at T-43 minutes. A third service arm is released at T-30 seconds, and a fourth at about T-16.5 seconds. The remaining five arms are set to swing back at vehicle first motion after T-0.

The service arms are equipped with a backup retraction system in case the primary mode fails.

The Apollo access arm (service arm 9), located at the 320-foot level above the launcher base, provides access to the spacecraft cabin for the closeout team and astronaut crews. The flight crew will board the spacecraft starting about T-2 hours, 40 minutes in the count. The access arm will be moved to a parked position, 12 degrees from the spacecraft, at about T-43 minutes. This is a distance of about three feet, which permits a rapid reconnection of the arm to the spacecraft in the event of an emergency condition. The arm is fully retracted at the T-5 minute mark in the count.

The Apollo 10 vehicle is secured to the mobile launcher by four combination support and hold-down arms mounted on the launcher deck. The hold-down arms are cast in one piece, about 6 x 9 feet at the base and 10 feet tall, weighing more than 20 tons. Damper struts secure the vehicle near its top.

After the engines ignite, the arms hold Apollo 10 for about six seconds until the engines build up to 95 percent thrust and other monitored systems indicate they are functioning properly. The arms release on receipt of a launch commit signal at the zero mark in the count. But the vehicle is prevented from accelerating too rapidly by controlled release mechanisms.

Transporter

The six-million-pound transporters, the largest tracked vehicles known, move mobile launchers into the VAB and mobile launchers with assembled Apollo space vehicles to the launch pad. They also are used to transfer the mobile service structure to and from the launch pads. Two transporters are in use at Complex 39.

The Transporter is 131 feet long and 114 feet wide. The vehicle moves on four double-tracked crawlers, each 10 feet high and 40 feet long. Each shoe on the crawler track is seven feet six inches in length and weighs about a ton.

Sixteen traction motors powered by four 1,000-kilowatt generators, which in turn are driven by two 2,750-horsepower diesel engines, provide the motive power for the transporter. Two 750-kw generators, driven by two 1,065-horsepower diesel engines, power the jacking, steering, lighting, ventilating and electronic systems.

Maximum speed of the transporter is about one-mile-per-hour loaded and about two-miles-per-hour unloaded. A five-mile trip to Pad B with a mobile launcher, made at less than maximum speed, takes approximately 10-12 hours.

The transporter has a leveling system designed to keep the top of the space vehicle vertical within plus-or-minus 10 minutes of arc -- about the dimensions of a basketball.

This system also provides leveling operations required to negotiate the five percent ramp which leads to the launch pad and keeps the load level when it is raised and lowered on pedestals both at the pad and within the VAB.

The overall height of the transporter is 20 feet from ground level to the top deck on which the mobile launcher is mated for transportation. The deck is flat and about the size of a base-ball diamond (90 by 90 feet).

Two operator control cabs, one at each end of the chassis located diagonally opposite each other, provide totally enclosed stations from which all operating and control functions are coordinated.

Crawlerway

The transporter moves on a roadway 131 feet wide, divided by a median strip. This is almost as broad as an eight-lane turnpike and is designed to accommodate a combined weight of about 18 million pounds.

The roadway is built in three layers with an average depth of seven feet. The roadway base layer is two-and-one-half feet of hydraulic fill compacted to 95 percent density. The next layer consists of three feet of crushed rock packed to maximum density, followed by a layer of one foot of selected hydraulic fill. The bed is topped and sealed with an asphalt prime coat.

On top of the three layers is a cover of river rock, eight inches deep on the curves and six inches deep on the straightway. This layer reduces the friction during steering and helps distribute the load on the transporter bearings.

Mobile Service Structure

A 402-foot-tall, 9.8-million-pound tower is used to service the Apollo launch vehicle and spacecraft at the pad. The 40-story steel-trussed tower, called a mobile service structure, provides 360-degree platform access to the Saturn launch vehicle and the Apollo spacecraft.

The service structure has five platforms -- two self-propelled and three fixed, but movable. Two elevators carry personnel and equipment between work platforms. The platforms can open and close around the 363-foot space vehicle.

After depositing the mobile launcher with its space vehicle on the pad, the transporter returns to a parking area about 13,000 feet from pad B. There it picks up the mobile service structure and moves it to the launch pad. At the pad, the huge tower is lowered and secured to four mount mechanisms.

The top three work platforms are located in fixed positions which serve the Apollo spacecraft. The two lower movable platforms serve the Saturn V.

The mobile service structure remains in position until about T-11 hours when it is removed from its mounts and returned to the parking area.

Water Deluge System

A water deluge system will provide a million gallons of industrial water for cooling and fire prevention during launch of Apollo 10. Once the service arms are retracted at liftoff, a spray system will come on to cool these arms from the heat of the five Saturn F-1 engines during liftoff.

On the deck of the mobile launcher are 29 water nozzles. This deck deluge will start immediately after liftoff and will pour across the face of the launcher for 30 seconds at the rate of 50,000 gallons-per-minute. After 30 seconds, the flow will be reduced to 20,000 gallons-per-minute.

Positioned on both sides of the flame trench are a series of nozzles which will begin pouring water at 8,000 gallons-per-minute, 10 seconds before liftoff. This water will be directed over the flame deflector.

Other flush mounted nozzles, positioned around the pad, will wash away any fluid spill as a protection against fire hazards.

Water spray systems also are available along the egress route that the astronauts and closeout crews would follow in case an emergency evacuation was required.

Flame Trench and Deflector

The flame trench is 58 feet wide and approximately six feet above mean sea level at the base. The height of the trench and deflector is approximately 42 feet.

The flame deflector weighs about 1.3 million pounds and is stored outside the flame trench on rails. Wehn it is moved beneath the launcher, it is raised hydraulically into position. The deflector is covered with a four-and-one-half-inch thickness of refractory concrete consisting of a volcanic ash aggregate and a calcuim aluminate binder. The heat and blast of the engines are expected to wear about three-quarters of an inch from this refractory surface during the Apollo 10 launch.

Pad Areas

Both Pad A and Pad B of Launch Complex 39 are roughly octagonal in shape and cover about one fourth of a square mile of terrain.

The center of the pad is a hardstand constructed of heavily reinforced concrete. In addition to supporting the weight of the mobile launcher and the Apollo Saturn V vehicle, it also must support the 9.8-million-pound mobile service structure and 6-million-pound transporter, all at the same time. The top of the pad stands some 48 feet above sea level.

Saturn V propellants -- liquid oxygen, liquid hydrogen and RP-1 -- are stored near the pad perimeter.

Stainless steel, vacuum-jacketed pipes carry the liquid oxygen (LOX) and liquid hydrogen from the storage tanks to the pad, up the mobile launcher, and finally into the launch vehicle propellant tanks.

LOX is supplied from a 900,000-gallon storage tank. A centrifugal pump with a discharge pressure of 320 pounds-per-square-inch pumps LOX to the vehicle at flow rates as high as 10,000-gallons-per-minute.

Liquid hydrogen, used in the second and third stages, is stored in an 850,000-gallon tank, and is sent through 1,500 feet of 10-inch, vacuum-jacketed invar pipe. A vaporizing heat exchanger pressurizes the storage tank to 60 psi for a 10,000 gallons-per-munute flow rate.

The RP-1 fuel, a high grade of kerosene is stored in three tanks--each with a capacity of 86,000 gallons. It is pumped at a rate of 2,000 gallons-per-minute at 175 psig.

The Complex 39 pneumatic system includes a converter-compressor facility, a pad high-pressure gas storage battety, a high-pressure storage battery in the VAB, low and high-pressure, cross-country supply lines, high-pressure hydrogen storage and conversion equipment, and pad distribution piping to pneumatic control panels. The various purging systems require 187,000 pounds of liquid nitrogen and 21,000 gallons of helium.

Pad B is virtually a twin of Pad A. The top of Pad B
is 5 feet higher in elevation above mean sea level than Pad
A to provide better drainage of the general area plus better
drainage from holding and burn ponds.

The electrical substation for Pad B is located under-
neath the west slope of the pad, whereas the corresponding
substation for Pad A is in the open approximately 150 feet
from the lower edge of the west slope of the pad. The pad
B design change was made to harden the substation against the
launch environment. The only other major difference is in
the location of the industrial/fire/potable water valve pit.
At Pad A, it's on the west side of the Pad and at Pad B it's
on the east side of the pad. The difference rests in the rout-
ing of water lines alongside the crawlerway.

Basic construction work on Pad B began on Dec. 7, 1964,
and the facility was accepted by the government on August 22,
1966. The intervening period has been spent in equipping the
pad and bringing it up to launch readiness.

Mission Control Center

The Mission Control Center at the Manned Spacecraft
Center, Houston, is the focal point for Apollo flight control
activities. The center receives tracking and telemetry data
from the Manned Space Flight Network, processes this data
through the Mission Control Center Real-Time Computer Complex,
and displays this data to the flight controllers and engineers
in the Mission Operations Control Room and staff support rooms.

The Manned Space Flight Network tracking and data
acquisition stations link the flight controllers at the center
to the spacecraft.

For Apollo 10 all network stations will be remote sites,
that is, without flight control teams. All uplink commands and
voice communications will originate from Houston, and telemetry
data will be sent back to Houston at high speed rates (2,400
bits-per-second), on two separate data lines. They can be
either real time or playback information.

Signal flow for voice circuits between Houston and
the remote sites is via commercial carrier, usually satellite,
wherever possible using leased lines which are part of the NASA
Communications Network.

Commands are sent from Houston to NASA's Goddard Space
Flight Center, Greenbelt, Md., on lines which link computers
at the two points. The Goddard communication computers provide automatic switching facilities and speed buffering for the
command data. Data are transferred from Goddard to remote sites
on high speed (2,400 bits-per-second) lines. Command loads also
can be sent by teletype from Houston to the remote sites at 100
words-per-minute. Again, Goddard computers provide storage and
switching functions.

-more-

Telemetry data at the remote site are received by the RF receivers, processed by the pulse code modulation ground stations, and transferred to the 642B remote-site telemetry computer for storage. Depending on the format selected by the telemetry controller at Houston, the 642B will send the desired format through a 2010 data transmission unit which provides parallel to serial conversion, and drives a 2,400 bit-per-second mode.

The data mode converts the digital serial data to phase-shifted keyed tones which are fed to the high speed data lines of the communications network.

Tracking data are sent from the sites in a low speed (100 words) teletype format and a 240-bit block high speed (2,400 bits) format. Data rates are one sample-6 seconds for teletype and 10 samples (frames) per second for high speed data.

All high-speed data, whether tracking or telemetry, which originate at a remote site are sent to Goddard on high-speed lines. Goddard reformats the data when necessary and sends them to Houston in 600-bit blocks at a 40,800 bits-per-second rate. Of the 600-bit block, 480 bits are reserved for data, the other 120 bits for address, sync, intercomputer instructions, and polynominal error encoding.

All wideband 40,800 bits-per-second data originating at Houston are converted to high speed (2,400 bits-per-second) data at Goddard before being transferred to the designated remote site.

-more-

MANNED SPACE FLIGHT NETWORK

The Manned Space Flight Network (MSFN) will support the complete Apollo spacecraft, operating at lunar distance, for the first time in Apollo 10. The network had its initial service with lunar distances in Apollo 8 last December, but that flight did not carry the lunar module.

For Apollo 10, the MSFN will employ 17 ground stations (including three wing, or backup, sites), four instrumented ships, and six to eight instrumented aircraft, to track spacecraft position and furnish a large volume of communications, television and telemetry services.

Essentially, the entire network is designed to provide reliable and continuous communications with the astronauts, launch vehicle and spacecraft from liftoff through lunar orbit to splashdown. It will keep ground controllers in close contact with the spacecraft and astronauts at all times, except for approximately 45 minutes when Apollo 10 will be behind the Moon during each lunar orbit and the time between stations while in Earth orbit.

As the space vehicle lifts off from Kennedy Space Center, the tracking stations will be watching it. As the Saturn ascends, voice and data will be instantaneously transmitted to Mission Control Center (MCC) in Houston. Data will be run through computers at MCC for visual display to flight controllers.

Depending on the launch azimuth, a string of 30-foot-diameter antennas around the Earth will keep tabs on Apollo 10 and transmit information back to Houston: beginning with the station at Merritt Island, Fla.; thence Grand Bahama Island, Bermuda; the tracking ship Vanguard; Canary Island; Carnarvon, Australia; Hawaii, tracking ship Redstone, Guaymas, Mexico; and Corpus Christi, Tex.

To inject Apollo 10 into translunar trajectory MCC will send a signal through one of the land stations or one of the Apollo ships in the Pacific. As the spacecraft head for the Moon, the engine burn will be monitored by the ships and an Apollo Range Instrumentation Aircraft (ARIA). The ARIA provides a relay for the astronauts' voices and data communication with Houston.

As the spacecraft moves away from Earth, the smaller 30-foot diameter antennas communicate first with the spacecraft. At a spacecraft altitude of 10,000 miles the tracking function goes to the more powerful 85-foot antennas. These are located near Madrid, Spain; Goldstone, Calif.; and Canberra, Australia.

MANNED SPACE FLIGHT TRACKING NETWORK

The 85-foot antennas are spaced at approximately 120-degree intervals around Earth so at least one antenna has the Moon in view at all times. As the Earth revolves from west to east, one station hands over control to the next station as it moves into view of the spacecraft. In this way, continuous data and communication flow is maintained.

Data are constantly relayed back through the huge antennas and transmitted via the NASA Communications Network (NASCOM) a half million miles of land and underseas cables and radio circuits, including those through communications satellites, to MCC. This information is fed into computers for visual display in Mission Control. For example, a display would show the exact position of the spacecraft on a large map. Returning data could indicate a drop in power or some other difficulty which would result in a red light going on to alert a flight controller to corrective action.

Returning data flowing to the Earth stations give the necessary information for commanding mid-course maneuvers to keep the Apollo 10 in a proper trajectory for orbiting the Moon. While the flight is in the vicinity of the Moon, these data indicate the amount of retrograde burn necessary for the service module engine to place the spacecraft units in lunar orbit.

Once the lunar module separates from the command module/service module and goes into a separate lunar orbit, the MSFN will be required to keep track of both craft and provide continuous two-way communications and telemetry between them and the Earth. The prime antenna at each of the three MSFN deep space tracking stations will handle one craft while the wing or back-up antenna at each of these stations will handle the other craft during each pass.

Continuous tracking and acquisition of data between Earth and the Apollo spacecraft will provide support for the Apollo rendezvous and docking maneuvers. This information also will be used to determine the time and duration of the service module propulsion engine burn required to place the command/service module into a precise trajectory for reentering the Earth's atmosphere at the planned location. As the spacecraft moves toward Earth at about 25,000 miles-per-hour, it must re-enter at the proper angle.

Data coming to the various tracking stations and ships are fed into the computers at MCC. From computer calculations, the flight controllers will provide the returning spacecraft with the necessary information to make an accurate reentry. Appropriate MSFN stations, including tracking ships and aircraft positioned in the Pacific for this event are on hand to provide support during reentry. An ARIA aircraft will relay astronaut voice communications to MCC and antennas on reentry ships will follow the spacecraft.

During the journey to the Moon and back, television will be received from the spacecraft at the three 85-foot antennas around the world, in Spain, California, and Australia. Scan converters permit immediate transmission of commercial quality television via NASCOM to Mission Control where it will be released to TV networks.

NASA Communications Network

The NASA Communications Network (NASCOM) consists of several systems of diversely routed communications channels leased on communications satellites, common carrier systems and high frequency radio facilities where necessary to provide the access links.

The system consists of both narrow and wide-band channels, and some TV channels. Included are a variety of telegraph, voice, and data systems (digital and analog) with several digital data rates. Wide-band systems do not extend overseas. Alternate routes or redundancy provide added reliability.

A primary switching center and intermediate switching and control points provide centralized facility and technical control, and switching operations under direct NASA control. The primary switching center is at the Goddard Space Flight Center, Greenbelt, Md. Intermediate switching centers are located at Canberra, Madrid, London, Honolulu, Guam, and Kennedy Space Center.

For Apollo 10, the Kennedy Space Center is connected directly to the Mission Control Center, Houston via the Apollo Launch Data System and to the Marshall Space Flight Center, Huntsville, Ala., by a Launch Information Exchange Facility. Both of these systems are part of NASCOM. They consist of data gathering and transmission facilities designed to handle launch data exclusively.

After launch, all network tracking and telemetry data hubs at GSFC for transmission to MCC Houton via two 50,000 bits-per-second circuits used for redundancy and in cast of data overflow.

Two Intelsat communications satellites will be used for Apollo 10. The Atlantic satellite will service the Ascension Island unified S-band (USB) station, the Atlantic Ocean ship and the Canary Islands site. These stations will be able to transmit through the satellite via the Comsat-operated ground station at Etam W.Va.

NASA COMMUNICATIONS NETWORK

The second Apollo Intelsat communications satellite over the mid-Pacific will service the Carnarvon, Australia USB site and the Pacific Ocean ships. All these stations will be able to transmit simultaneously through the satellite to Houston via Brewster Flat, Wash., and the Goddard Space Flight Center, Greenbelt, Md.

Network Computers

At fraction-of-a-second intervals, the network's digital data processing systems, with NASA's Manned Spacecraft Center as the focal point, "talk" to each other or to the spacecraft. High-speed computers at the remote site (tracking ships included) issue commands or "up-link" data on such matters as control of cabin pressure, orbital guidance commands, or "go-no-go" indications to perform certain functions.

When information originates from Houston, the computers refer to their pre-programmed information for validity before transmitting the required data to the spacecraft.

Such "up-link" information is comminicated by ultra-high-frequency radio about 1,200 bits-per-second. Communication between remote ground sites, via high-speed communications links, occurs at about the same rate. Houston reads information from these ground sties at 2,400 bits-per-second, as well as from remote sites at 100 words-per-minute.

The computer systems perform many other functions, including:

. Assuring the quality of the transmission lines by continually exercising data paths.

. Verifying accuracy of the messages by repetitive operations.

. Constantly updating the flight status.

For "down link" data, sensors built into the spacecraft continually sample cabin temperature, pressure, physical information on the astronauts such as heartbeat and respiration, among other items. These data are transmitted to the ground stations at 51.2 kilobits (12,800 binary digits) per-second.

At MCC the computers:

. Detect and select changes or deviations, compare with their stored programs, and indicate the problem areas or pertinent data to the flight controllers.

. Provide displays to mission personnel.

. Assemble output data in proper formats.

. Log data on magnetic tape for replay for the flight
  controllers.

. Keep time.

The Apollo Ships

The mission will be supported by four Apollo instrumentation ships operating as integral stations of the Manned Space
Flight Network (MSFN) to provide coverage in areas beyond
the range of land stations.

The ships, Vanguard, Redstone, Mercury, and Huntsville,
will perform tracking, telemetry, and communication functions
for the launch phase, Earth orbit insertion, translunar injection and reentry at the end of the mission.

Vanguard will be stationed about 1,000 miles southeast
of Bermuda (25 degrees N, 49 degrees W) to bridge the Bermuda-
Antigua gap during Earth orbit insertion. Vanguard also function
as part of the Atlantic recovery fleet in the event of a launch
phase contingency. The Redstone, at 14 degrees S, 145.5 degrees
E; Mercury, 32 degrees S, 131 degrees E; and Huntsville, 17 degrees S. 174 degrees W, provide a triangle of mobile stations between the MSFN stations at Carnarvon and Hawaii for coverage of
the burn interval for translunar injection. In the event the
launch data slips from May 18, the ships will all move generally
northeastward to cover the changing flight window patterns.

Redstone and Huntsville will be repositioned along the
reentry corridor for tracking, telemetry, and communications
functions during reentry and landing. They will track Apollo
from about 1,000 miles away through communications blackout
when the spacecraft will drop below the horizon and will be
picked up by the ARIA aircraft.

The Apollo ships were developed jointly by NASA and the
Department of Defense. The DOD operates the ships in support
of Apollo and other NASA and DOD missions on a non-interference
basis with Apollo requirements.

Management of the Apollo ships is the responsibility
of the Commander, Air Force Western Test Range (AFWTR). The
Military Sea Transport Service provides the maritime crews and
the Federal Electric Corp., International Telephone and Telegraph, under contract to AFWTR, provides the technical instrumentation crews.

The technical crews operate in accordance with joint
NASA/DOD standards and specifications which are compatible
with MSFN operational procedures.

Apollo Range Instrumentation Aircraft (ARIA)

The Apollo Range Instrumentation Aircraft will support
the mission by filling gaps in both land and ship station
coverage where important and significant coverage requirements
exist.

During Apollo 10, the ARIA will be used primarily to
fill coverage gaps of the land and ship stations in the Indian
Ocean and in the Pacific between Australia and Hawaii during
the translunar injection interval. Prior to and during the
burn, the ARIA record telemetry data from Apollo provide a
real-time voice communication between the astronauts and the
flight director at Houston.

Eight aircraft will participate in this mission, operating
from Pacific, Australian and Indian Ocean air fields in
positions under the orbital track of the spacecraft and booster.
The aircraft, like the tracking ships, will be redeployed in a
northeastward direction in the event of launch day slips.

For reentry, the ARIA will be redeployed to the landing
area to continue communications between Apollo and Mission
Control and provide position information on the spacecraft
after the blackout phase of reentry has passed.

The total ARIA fleet for Apollo missions consist of
eight EC-135A (Boeing 707) jets equipped specifically to
meet mission needs. Seven-foot parabolic antennas have been
installed in the nose section of the planes giving them a
large, bulbous look.

The aircraft, as well as flight and instrumentation
crews, are provided by the Air Force and they are equipped
through joint Air Force-NASA contract action. ARIA operate
in Apollo missions in accordance with MSFN procedures.

Ship Positions for Apollo 10

May 18, 1969

Insertion Ship (VAN)	25 degrees N - 49 degrees W
Injection Ship (MER)	32 degrees S - 131 degrees E
Injection Ship (RED)	14 degrees S - 145.5 degrees E
Injection Ship (RED)	20 degrees S - 172.5 degrees E
Reentry Support	
Reentry Ship (HTV)	17 degrees S - 174 degrees W

May 20, 1969

Insertion Ship (VAN)	25 degrees N - 49 degrees W
Injection Ship (MER)	32 degrees S - 131 degrees E
Injection Ship (RED)	14 degrees S - 145.5 degrees E
Injection ship (RED)	13 degrees S - 174 degrees E
Reentry Support	
Reentry Ship (HTV)	8 degrees S - 173 degrees W

May 23, 1969

Insertion Ship (VAN)	25 degrees N - 49 degrees W
Injection Ship (MER)	Released
Injection Ship (RED)	7.5 degrees S - 156 degrees E
Injection Ship (RED)	1 degree N - 177.5 degrees E
Reentry Support	
Reentry Ship (HTV)	10 degrees N - 172 degrees W

May 24, 1969

Insertion Ship (VAN)	25 degrees N - 49 degrees W
Injection Ship (MER)	Released
Injection Ship (RED)	3 degrees S - 158 degrees E
Injection Ship (RED)	9 degrees N - 175.5 degrees E
Reentry Support	
Reentry Ship (HTV)	15.5 degrees N - 173 degrees W

May 25, 1969

Insertion Ship (VAN)	25 degrees N - 49 degrees W
Injection Ship (MER)	Released
Injection Ship (RED)	0.5 degrees N - 161 degrees E
Injection Ship (RED)	16 degrees N - 174 degrees E
Reentry Support	
Reentry Ship (HTV)	22 degrees N - 173 degrees W

-more-

APOLLO PROGRAM MANAGEMENT

The Apollo Program, the United States' effort to land men on the Moon and return them safely to Earth before 1970, is the responsibility of the Office of Manned Space Flight (OMSF), National Aeronautics and Space Administration, Washington, D.C. Dr. George E. Mueller is Associate Administrator for Manned Space Flight.

NASA Manned Spacecraft Center (MSC), Houston, is responsible for development of the Apollo spacecraft, flight crew training and flight control. Dr. Robert R. Gilruth is Center Director.

NASA Marshall Space Flight Center (MSFC), Huntsville, Ala., is responsible for development of the Saturn launch vehicles. Dr. Wernher von Braun is Center Director.

NASA John F. Kennedy Space Center (KSC), Fla., is responsible for Apollo/Saturn launch operations. Dr. Kurt H. Debus is Center Director.

NASA Goddard Space Flight Center (GSFC), Greenbelt, Md., manages the Manned Space Flight Network under the direction of the NASA Office of Tracking and Data Acquisition (OTDA). Gerald M. Truszynski is Associate Administrator for Tracking and Data Acquisition. Dr. John F. Clark is Director of GSFC.

Apollo/Saturn Officials

NASA HEADQUARTERS

Lt. Gen. Sam C. Phillips, (USAF)   Apollo Program Director, OMSF

George H. Hage   Apollo Program Deputy Director, Mission Director, OMSF

Chester M. Lee   Assistant Mission Director, OMSF

Col. Thomas H. McMullen (USAF)   Assistant Mission Director, OMSF

Maj. Gen. James W. Humphreys, Jr.   Director of Space Medicine, OMSF

Norman Pozinsky   Director, Network Support Implementation Div., OTDA

-more-

Manned Spacecraft Center

George M. Low	Manager, Apollo Spacecraft Program
Kenneth S. Kleinknecht	Manager, Command and Service Modules
Brig. Gen. C. H. Bolender (USAF)	Manager, Lunar Module
Donald K. Slayton	Director of Flight Crew Operations
Christopher C. Kraft, Jr.	Director of Flight Operations
Glynn S. Lunney	Flight Director
Milton L. Windler	Flight Director
M. P. Frank	Flight Director
Gerald Griffin	Flight Director
Charles A. Berry	Director of Medical Research and Operations

Marshall Space Flight Center

Maj. Gen. Edmund F. O'Connor	Director of Industrial Operations
Dr. F. A. Speer	Director of Mission Operations
Lee B. James	Manager, Saturn V Program Office
William D. Brown	Manager, Engine Program Office

Kennedy Space Center

Miles Ross	Deputy Director, Center Operations
Rocco A. Petrone	Director, Launch Operations
Raymond L. Clark	Director, Technical Support
Rear Adm. Roderick O. Middleton (USN)	Manager, Apollo Program Office
Walter J. Kapryan	Deputy Director, Launch Operations
Dr. Hans F. Gruene	Director, Launch Vehicle Operations
John J. Williams	Director, Spacecraft Operations

Paul C. Donnelly   Launch Operations Manager

Goddard Space Flight Center

Ozro M. Covington   Assistant Director for Manned
Space Flight Tracking

Henry F. Thompson   Deputy Assistant Director for
Manned Space Flight Support

H. William Wood   Chief, Manned Flight Operations
Div.

Tecwyn Roberts   Chief. Manned Flight Engineering
Div.

Department of Defense

Maj. Gen. Vincent G. Huston, (USAF)   DOD Manager of Manned Space
Flight Support Operations

Maj. Gen. David M. Jones, (USAF)   Deputy DOD Manager of Manned
Space Flight Support Oper-
ations, Commander of USAF
Eastern Test Range

Rear Adm. F. E. Bakutis, (USN)   Commander of Combined Task Force
130, Pacific Recovery Area

Rear Adm. P. S. McManus, (USN)   Commander of Combined Task Force
140, Atlantic Recovery Area

Col. Royce G. Olson, (USAF)   Director of DOD Manned Space
Flight Office

Brig. Gen. Allison C. Brooks,
(USAF)   Commander Aerospace Rescue and
Recovery Service

Major Apollo/Saturn V Contractors

Contractor	Item
Bellcomm Washington, D.C.	Apollo Systems Engineering
The Boeing Co. Washington, D.C.	Technical Integration and Evaluation
General Electric-Apollo Support Dept., Daytona Beach, Fla.	Apollo Checkout, and Quality and Reliability
North American Rockwell Corp. Space Div., Downey, Calif.	Command and Service Modules
Grumman Aircraft Engineering Corp., Bethpage, N.Y.	Lunar Module
Massachusetts Institute of Technology, Cambridge, Mass.	Guidance & Navigation (Technical Management)
General Motors Corp., AC Electronics Div., Milwaukee, Wis.	Guidance & Navigation (Manufacturing)
TRW Systems Inc. Redondo Beach, Calif.	Trajectory Analysis
Avco Corp., Space Systems Div., Lowell, Mass.	Heat Shield Ablative Material
North American Rockwell Corp. Rocketdyne Div. Canoga Park, Calif.	J-2 Engines, F-1 Engines
The Boeing Co. New Orleans	First Stage (SIC) of Saturn V Launch Vehicles, Saturn V Systems Engineering and Integration, Ground Support Equipment
North American Rockwell Corp. Space Div. Seal Beach, Calif.	Development and Production of Saturn V Second Stage (S-II)
McDonnell Douglas Astronautics Co. Huntington Beach, Calif.	Development and Production of Saturn V. Third Stage (S-IVB)

-more-

International Business Machines Federal Systems Div. Huntsville, Ala.	Instrument Unit
Bendix Corp. Navigation and Control Div. Teterboro, N.J.	Guidance Components for Instrument Unit (Including ST-124M Stabilized Platform)
Federal Electric Corp.	Communications and Instrumentation Support, KSC
Bendix Field Engineering Corp.	Launch Operations/Complex Support, KSC
Catalytic-Dow	Facilities Engineering and Modifications, KSC
Hamilton Standard Division United Aircraft Corp. Windsor Locks, Conn.	Portable Life Support System; LM ECS
ILC Industries Dover, Del.	Space Suits
Radio Corp. of America Van Nuys, Calif.	110A Computer - Saturn Checkout
Sanders Associates Nashua, N.H.	Operational Display Systems Saturn
Brown Engineering Huntsville, Ala.	Discrete Controls
Reynolds, Smith and Hill Jacksonville, Fla.	Engineering Design of Mobile Launchers
Ingalls Iron Works Birmingham, Ala.	Mobile Launchers (ML) (structural work)
Smith/Ernst (Joint Venture) Tampa, Fla. Washington, D.C.	Electrical Mechanical Portion of MLs
Power Shovel, Inc. Marion, Ohio	Transporter
Hayes International Birmingham, Ala.	Mobile Launcher Service Arms

APOLLO GLOSSARY

Ablating Materials--Special heat-dissipating materials on the surface of a spacecraft that vaporize during reentry.

Abort--The unscheduled termination of a mission prior to its completion.

Accelerometer--An instrument to sense accelerative forces and convert them into corresponding electrical quantities usually for controlling, measuring, indicating or recording purposes.

Adapter Skirt--A flange or extension of a stage or section that provides a ready means of fitting another stage or section to it.

Antipode--Point on surface of planet exactly 180 degrees opposite from reciprocal point on a line projected through center of body. In Apollo usage, antipode refers to a line from the center of the Moon through the center of the Earth and pro- jected to the Earth surface on the opposite side. The anti- pode crosses the mid-Pacific recovery line along the 165th meridian of longitude once each 24 hours.

Apocynthion--Point at which object in lunar orbit is farthest from the lunar surface -- object having been launched from body other than Moon. (Cynthia, Roman goddess of Moon)

Apogee--The point at which a Moon or artificial satellite in its orbit is farthest from Earth.

Apolune--Point at which object launched from the Moon into lunar orbit is farthest from lunar surface, e.g.: ascent stage of lunar module after staging into lunar orbit following lunar landing.

Attitude--The position of an aerospace vehicle as determined by the inclination of its axes to some frame of reference; for Apollo, an inertial, space-fixed reference is used.

Burnout--The point when combustion ceases in a rocket engine.

Canard--A short, stubby wing-like element affixed to the launch escape tower to provide CM blunt end forward aerodynamic capture during an abort.

Celestial Guidance--The guidance of a vehicle by reference to celestial bodies.

-more-

Celestial Mechanics--The science that deals primarily with the
    effect of force as an agent in determining the orbital
    paths of celestial bodies.

Cislunar--Adjective referring to space between Earth and the Moon,
    or between Earth and Moon's orbit.

Closed Loop--Automatic control units linked together with a
    process to form an endless chain.

Deboost--A retrograde maneuver which lowers either perigee or
    apogee of an orbiting spacecraft. Not to be confused with
    deorbit.

Declination--Angular measurement of a body above or below celestial
    equator, measured north or south along the body's hour
    circle. Corresponds to Earth surface latitude.

Delta V--Velocity change.

Digital Computer--A computer in which quantities are represented
    numerically and which can be used to solve complex problems.

Down-Link--The part of a communication system that receives, pro-
    cesses and displays data from a spacecraft.

Entry Corridor--The final flight path of the spacecraft before
    and during Earth reentry.

Ephemeris--Orbital measurements (apogee, perigee, inclination,
    period, etc.) of one celestial body in relation to another
    at given times. In spaceflight, the orbital measurements
    of a spacecraft relative to the celestial body about which
    it orbited.

Escape Velocity--The speed a body must attain to overcome a
    gravitational field, such as that of Earth; the velocity
    of escape at the Earth's surface is 36,700 feet-per-second.

Explosive Bolts--Bolts destroyed or severed by a surrounding
    explosive charge which can be activated by an electrical
    impulse.

Fairing--A piece, part or structure having a smooth, stream-
    lined outline, used to cover a nonstreamlined object or to
    smooth a junction.

Flight Control System--A system that serves to maintain attitude
    stability and control during flight.

Fuel Cell--An electrochemical generator in which the chemical
energy from the reaction of oxygen and a fuel is converted directly into electricity.

g or g Force--Force exerted upon an object by gravity or by
reaction to acceleration or deceleration, as in a change
of direction: one g is the measure of force required to
accelerate a body at the rate of 32.16 feet-per-second.

Gimbaled Motor--A rocket motor mounted on gimbal; i.e.: on a
contrivance having two mutually perpendicular axes of rotation, so as to obtain pitching and yawing correction moments.

Guidance System--A system which measures and evaluates flight
information, correlates this with target data, converts
the result into the conditions necessary to achieve the
desired flight path, and communicates this data in the form
of commands to the flight control system.

Heliocentric--Sun-centered orbit or other activity which has the
Sun at its center.

Inertial Guidance--Guidance by means of the measurement and
integration of acceleration from on board the spacecraft.
A sophisticated automatic navigation system using gyro-
scopic devices, accelerometers etc., for high-speed vehicles.
It absorbs and interprets such data as speed, position, etc.,
and automatically adjusts the vehicle to a pre-determined
flight path. Essentially, it knows where it's going and
where it is by knowing where it came from and how it got
there. It does not give out any radio frequency signal so
it cannot be detected by radar or jammed.

Injection--The process of boosting a spacecraft into a calcu-
lated trajectory.

Insertion--The process of boosting a spacecraft into an orbit
around the Earth or other celestial bodies.

Multiplexing--The simultaneous transmission of two or more signals within a single channel. The three basic methods
of multiplexing involve the separation of signals by time
division, frequency division and phase division.

Optical Navigation--Navigation by sight, as opposed to inertial
methods, using stars or other visible objects as reference.

Oxidizer--In a rocket propellant, a substance such as liquid
oxygen or nitrogen tetroxide which supports combustion of
the fuel.

Penumbra--Semi-dark portion of a shadow in which light is partly cut off, e.g.: surface of Moon or Earth away from Sun where the disc of the Sun is only partly obscured.

Pericynthion--Point nearest Moon of object in lunar orbit--object having been launched from body other than Moon.

Perigee--Point at which a Moon or an artificial satellite in its orbit is closest to the Earth.

Perilune--The point at which a satellite (e.g.: a spacecraft) in its orbit is closest to the Moon. Differs from pericynthion in that the orbit is Moon-originated.

Pitch--The movement of a space vehicle about an axis (Y) that is perpendicular to its longitudinal axis.

Reentry--The return of a spacecraft that reenters the atmosphere after flight above it.

Retrorocket--A rocket that gives thrust in a direction opposite to the direction of the object's motion.

Right Ascension--Angular measurement of a body eastward along the celestial equator from the vernal equinox (0 degrees RA) to the hour circle of the body. Corresponds roughly to Earth surface longitude, except as expressed in hrs:min:sec instead of 180 degrees west and east from 0 degrees (24 hours=360 degrees).

Roll--The movements of a space vehicle about its longitudinal (X) axis.

S-Band--A radio-frequency band of 1,550 to 5,200 megahertz.

Selenographic--Adjective relating to physical geography of Moon. Specifically, positions on lunar surface as measured in latitude from lunar equator and in longitude from a reference lunar meridian.

Selenocentric--Adjective referring to orbit having Moon as center. (Selene, Gr. Moon)

Sidereal--Adjective relating to measurement of time, position or angle in relation to the celestial sphere and the vernal equinox.

State vector--Ground-generated spacecraft position, velocity and timing information uplinked to the spacecraft computer for crew use as a navigational reference.

Telemetering--A system for taking measurements within an aero-
    space vehicle in flight and transmitting them by radio to
    a ground station.

Terminator--Separation line between lighted and dark portions
    of celestial body which is not self luminous.

Ullage--The volume in a closed tank or container that is not
    occupied by the stored liquid; the ratio of this volume
    to the total volume of the tank; also an acceleration to
    force propellants into the engine pump intake lines before
    ignition.

Umbra--Darkest part of a shadow in which light is completely
    absent, e.g.: surface of Moon or Earth away from Sun where
    the disc of the Sun is completely obscured.

Update pad--Information on spacecraft attitudes, thrust values,
    event times, navigational data, etc., voiced up to the crew
    in standard formats according to the purpose, e.g.: maneuver
    update, navigation check, landmark tracking, entry update,
    etc.

Up-Link Data--Information fed by radio signal from the ground to
    a spacecraft.

Yaw--Angular displacement of a space vehicle about its vertical
    (Z) axis.

APOLLO ACRONYMS AND ABBREVIATIONS

(Note: This list makes no attempt to include all Apollo
program acronyms and abbreviations, but several are listed
that will be encountered frequently in the Apollo 10 mission.
Where pronounced as words in air-to-ground transmissions,
acronyms are phonetically shown in parentheses. Otherwise,
abbreviations are sounded out by letter.)

AGS	(Aggs)	Abort Guidance System (LM)
AK		Apogee kick
APS	(Apps)	Ascent Propulsion System (LM)
		Auxiliary Propulsion System (S-IVB stage)
BMAG	(Bee-mag)	Body mounted attitude gyro
CDH		Constant delta height
CMC		Command Module Computer
COI		Contingency orbit insertion
CRS		Concentric rendezvous sequence
CSI		Concentric sequence initiate
DAP	(Dapp)	Digital autopilot
DEDA	(Dee-da)	Data Entry and Display Assembly (LM AGS)
DFI		Development flight instrumentation
DOI		Descent orbit insertion
DPS	(Dips)	Descent propulsion system
DSKY	(Diskey)	Display and keyboard
EPO		Earth Parking Orbit
FDAI		Flight director attitude indicator
FITH	(Fith)	Fire in the hole (LM ascent abort staging)
FTP		Fixed throttle point
HGA		High-gain antenna
IMU		Inertial measurement unit

IRIG	(Ear-ig)	Inertial rate integrating gyro
LOI		Lunar orbit insertion
LPO		Lunar parking orbit
MCC		Mission Control Center
MC&W		Master caution and warning
MSI		Moon sphere of influence
MTVC		Manual thrust vector control
NCC		Combined corrective maneuver
NSR		Coelliptical maneuver
PIPA	(Pippa)	Pulse integrating pendulous accelerometer
PLSS	(Pliss)	Portable life support system
PTC		Passive thermal control
PUGS	(Pugs)	Propellant utilization and gaging system
REFSMMAT	(Refsmat)	Reference to stable member matrix
RHC		Rotation hand controller
RTC		Real-time command
SCS		Stabilization and control system
SLA	(Slah)	Spacecraft LM adapter
SPS		Service propulsion system
TEI		Transearth injection
THC		Thrust hand controller
TLI		Translunar injection
TPF		Terminal phase finalization
TPI		Terminal phase initiate
TVC		Thrust vector control

-more-

CONVERSION FACTORS

	Multiply	By	To Obtain
Distance:			
	feet	0.3048	meters
	meters	3.281	feet
	kilometers	3281	feet
	kilometers	0.6214	statute miles
	statute miles	1.609	kilometers
	nautical miles	1.852	kilometers
	nautical miles	1.1508	statute miles
	statute miles	0.86898	nautical miles
	statute mile	1760	yards
Velocity:			
	feet/sec	0.3048	meters/sec
	meters/sec	3.281	feet/sec
	meters/sec	2.237	statute mph
	feet/sec	0.6818	statute miles/hr
	feet/sec	0.5925	nautical miles/hr
	statute miles/hr	1.609	km/hr
	nautical miles/hr (knots)	1.852	km/hr
	km/hr	0.6214	statute miles/hr
Liquid measure, weight:			
	gallons	3.785	liters
	liters	0.2642	gallons
	pounds	0.4536	kilograms
	kilograms	2.205	pounds

- more -

Multiply	By	To Obtain
Volume:		
cubic feet	0.02832	cubic meters
Pressure:		
pounds/sq inch	70.31	grams/sq cm

-end-

www.ingramcontent.com/pod-product-compliance
Lightning Source LLC
Chambersburg PA
CBHW051218200326
41519CB00025B/7158